HARCOURT Science

Harcourt School Publishers

Orlando • Boston • Dallas • Chicago • San Diego

www.harcourtschool.com

The **panther chameleon** (*Chamaeleo pardalis*) is native to the eastern and northern coasts of Madagascar and some surrounding islands. The panther chameleon lives in hot, humid rain forests and it eats insects. It can grow to be as long as 30 cm (about 1 ft). It feeds by capturing insects with a sticky tongue that can be as long as its entire body. The tongue can extend out and capture an insect in less than 1/16th of a second. The inside covers of this book show a closeup of the skin of the panther chameleon.

Copyright © by Harcourt, Inc.
2005 Edition

All rights reserved. No part of this publication may be reproduced or transmitted in any form or by any means, electronic or mechanical, including photocopy, recording, or any information storage and retrieval system, without permission in writing from the publisher.

Requests for permission to make copies of any part of the work should be addressed to School Permissions and Copyrights, Harcourt, Inc., 6277 Sea Harbor Drive, Orlando, Florida 32887-6777. Fax 407-345-2418.

HARCOURT and the Harcourt Logo are trademarks of Harcourt, Inc., registered in the United States of America and/or other jurisdictions.

sciLINKS is owned and provided by the National Science Teachers Association. All rights reserved.

Smithsonian Institution Internet Connections owned and provided by the Smithsonian Institution. All other material owned and provided by Harcourt School Publishers under copyright appearing above.

The name of the Smithsonian Institution and the Sunburst logo are registered trademarks of the Smithsonian Institution. The copyright in the Smithsonian website and Smithsonian website pages are owned by the Smithsonian Institution.

Printed in the United States of America

ISBN 0-15-325389-4   UNIT A
ISBN 0-15-325390-8   UNIT B
ISBN 0-15-325391-6   UNIT C
ISBN 0-15-325392-4   UNIT D
ISBN 0-15-325393-2   UNIT E
ISBN 0-15-325394-0   UNIT F

7 8 9 10 11 12 13 032 10 09 08 07 06 05 04

Authors

Marjorie Slavick Frank
Former Adjunct Faculty Member
Hunter, Brooklyn, and
    Manhattan Colleges
New York, New York

Robert M. Jones
Professor of Education
University of Houston–
    Clear Lake
Houston, Texas

Gerald H. Krockover
*Professor of Earth and Atmospheric
    Science Education*
School Mathematics and
    Science Center
Purdue University
West Lafayette, Indiana

Mozell P. Lang
Science Education Consultant
Michigan Department
    of Education
Lansing, Michigan

Joyce C. McLeod
Visiting Professor
Rollins College
Winter Park, Florida

Carol J. Valenta
*Vice President—Education, Exhibits,
    and Programs*
St. Louis Science Center
St. Louis, Missouri

Barry A. Van Deman
*Program Director, Informal Science
    Education*
Arlington, Virginia

Senior Editorial Advisor

Napoleon Adebola Bryant, Jr.
Professor Emeritus of Education
Xavier University
Cincinnati, Ohio

Program Advisors

Michael J. Bell
Assistant Professor of Early Childhood Education
School of Education
West Chester University
West Chester, Pennsylvania

George W. Bright
Professor of Mathematics Education
The University of North Carolina at Greensboro
Greensboro, North Carolina

Pansy Cowder
Science Specialist
Tampa, Florida

Robert H. Fronk
Head, Science/Mathematics Education Department
Florida Institute of Technology
Melbourne, Florida

Bernard A. Harris, Jr.
Physician and Former Astronaut
(STS 55–Space Shuttle *Columbia*, STS 63–Space Shuttle *Discovery*)
President, The Harris Foundation
Houston, Texas

Lois Harrison-Jones
Education and Management Consultant
Dallas, Texas

Kenneth R. Mechling
Professor of Biology and Science Education
Clarion University of Pennsylvania
Clarion, Pennsylvania

Nancy Roser
Professor of Language and Literacy Studies
University of Texas, Austin
Austin, Texas

Program Advisor and Activities Writer

Barbara ten Brink
Science Director
Round Rock Independent School District
Round Rock, Texas

Reviewers and Contributors

Rose G. Baublitz
Director, Curriculum and Staff Development
Granville Exempted Village School District
Granville, Ohio

Sharon W. Bowers
Teacher
Kemps Landing Magnet School
Virginia Beach, Virginia

Joe E. Hart
Science Lead Teacher
Clayton County Public Schools
Morrow, Georgia

Kathryn Henry
Teacher Trainer
Public School CS 200
New York, New York

Denise Hunt
Teacher
Cumming Elementary School
Cumming, Georgia

Marie A. McDermott
Teacher
Kingsbury School
Waterbury, Connecticut

Clyde Partner
Science/Health Curriculum Coordinator
Evanston S.D. #65
Evanston, Illinois

Michael F. Ryan
Educational Technology Specialist
Lake County Schools
Tavares, Florida

Arnold E. Serotsky
Teacher/Science Department Coordinator
Greece Athena Middle School
Rochester, New York

Lynda Wood
Science Coordinator
Southfield Education Center
Southfield, Michigan

UNIT A — LIFE SCIENCE
Living Systems

Unit Experiment	**A1**

CHAPTER 1 — From Single Cells to Body Systems — A2
Lesson 1—What Are Cells, and What Do They Do? A4
Lesson 2—How Do Body Systems Transport Materials? A14
Lesson 3—How Do Bones, Muscles, and Nerves Work Together? A22
    Science and Technology • Potato Vaccines A28
    People in Science • Bernard A. Harris, Jr. A30
    Activities for Home or School A31
Chapter Review and Test Preparation A32

CHAPTER 2 — Classifying Living Things — A34
Lesson 1—How Do Scientists Classify Living Things? A36
Lesson 2—How Are Animals Classified? A42
Lesson 3—How Are Plants Classified? A48
    Science Through Time • Naming Living Things A54
    People in Science • Ynes Enriquetta Julietta Mexia A56
    Activities for Home or School A57
Chapter Review and Test Preparation A58

CHAPTER 3 — Animal Growth and Heredity — A60
Lesson 1—How Do Animals Grow and Reproduce? A62
Lesson 2—What Is a Life Cycle? A70
Lesson 3—Why Are Offspring Like Their Parents? A76
    Science and Technology • Bionic Dog A82
    People in Science • Eduardo S. Cantu A84
    Activities for Home or School A85
Chapter Review and Test Preparation A86

CHAPTER 4 — Plants and Their Adaptations — A88
Lesson 1—What Are the Functions of Roots, Stems, and Leaves? A90
Lesson 2—How Do Plants Reproduce? A98
Lesson 3—How Do People Use Plants? A108
    Science and Technology • Corn Cards and Super Slurpers A114
    People in Science • Shirley Mah Kooyman A116
    Activities for Home or School A117
Chapter Review and Test Preparation A118

Unit Expeditions   **A120**

UNIT B — LIFE SCIENCE
Systems and Interactions in Nature

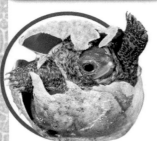

Unit Experiment	**B1**

CHAPTER 1 — Cycles In Nature — B2
Lesson 1—How Does Nature Reuse Materials? B4
Lesson 2—Why Is the Water Cycle Important? B12
    Science and Technology • Wetlands with a Purpose B18
    People in Science • Marjory Stoneman Douglas B20
    Activities for Home or School B21
Chapter Review and Test Preparation B22

CHAPTER 2 — Living Things Interact — B24
Lesson 1—What Are Ecosystems? B26
Lesson 2—How Does Energy Flow Through an Ecosystem? B32
Lesson 3—How Do Organisms Compete and Survive in an Ecosystem? ...B40
Lesson 4—What Is Extinction and What Are Its Causes? B48
    Science and Technology • The New Zoos B54
    People in Science • Dorothy McClendon B56
    Activities for Home or School B57
Chapter Review and Test Preparation B58

CHAPTER 3 — Biomes — B60
Lesson 1—What Are Land Biomes? B62
Lesson 2—What Are Water Ecosystems? B74
    Science and Technology • "See" Food B82
    People in Science • Alissa J. Arp B84
    Activities for Home or School B85
Chapter Review and Test Preparation B86

CHAPTER 4 — Protecting and Preserving Ecosystems — B88
Lesson 1—How Do Ecosystems Change Naturally? B90
Lesson 2—How Do People Change Ecosystems? B96
Lesson 3—How Can People Treat Ecosystems More Wisely? B102
Lesson 4—How Can People Help Restore Damaged Ecosystems? B108
    Science Through Time • Major Events in Environmental Awareness ..B114
    People in Science • Raman Sukumar B116
    Activities for Home or School B117
Chapter Review and Test Preparation B118

Unit Expeditions	**B120**

UNIT C — EARTH SCIENCE
Processes That Change the Earth

Unit Experiment	**C1**

CHAPTER 1 — Changes to Earth's Surface — C2
Lesson 1—What Processes Change Landforms? C4
Lesson 2—What Causes Mountains, Volcanoes, and Earthquakes? C12
Lesson 3—How Has Earth's Surface Changed? C20
    Science and Technology • Exploring Earth's Surface from Space C26
    People in Science • Kia K. Baptist C28
    Activities for Home or School C29
Chapter Review and Test Preparation C30

CHAPTER 2 — Rocks and Minerals — C32
Lesson 1—What Are Minerals? C34
Lesson 2—What Are Rocks? C40
Lesson 3—What Is the Rock Cycle? C48
    Science and Technology • Diamond Coatings C54
    People in Science • Mack Gipson, Jr. C56
    Activities for Home or School C57
Chapter Review and Test Preparation C58

CHAPTER 3 — Weather and Climate — C60
Lesson 1—How Can You Observe and Measure Weather Conditions? C62
Lesson 2—What Causes Weather? C70
Lesson 3—What Is Climate and How Does It Change? C78
    Science Through Time • Major Events in Weather Forecasting C86
    People in Science • Carolyn Kloth C88
    Activities for Home or School C89
Chapter Review and Test Preparation C90

CHAPTER 4 — Exploring the Oceans — C92
Lesson 1—What Are the Oceans Like? C94
Lesson 2—How Do Ocean Waters Move? C100
Lesson 3—How Do Oceans Interact with the Land? C108
Lesson 4—How Do People Explore the Oceans and Use Ocean
          Resources? C114
    Science and Technology • Saltwater Agriculture C122
    People in Science • Robert D. Ballard C124
    Activities for Home or School C125
Chapter Review and Test Preparation C126

Unit Expeditions — **C128**

UNIT D — EARTH SCIENCE
The Solar System and Beyond

Unit Experiment	**D1**

CHAPTER 1 — Earth, Moon, and Beyond — D2

Lesson 1—How Do the Earth and the Moon Compare? D4
Lesson 2—What Else Is in the Solar System? D12
Lesson 3—How Have People Explored the Solar System? D20
    Science Through Time • The History of Rockets and Spaceflight D28
    People in Science • Harrison Schmitt D30
    Activities for Home or School D31
Chapter Review and Test Preparation D32

CHAPTER 2 — The Sun and Other Stars — D34

Lesson 1—What Are the Features of the Sun? D36
Lesson 2—How Are Stars Classified? D44
Lesson 3—What Are Galaxies? D52
    Science and Technology • Magnetars D58
    People in Science • Julio Navarro D60
    Activities for Home or School D61
Chapter Review and Test Preparation D62

Unit Expeditions   **D64**

UNIT E — PHYSICAL SCIENCE
Building Blocks of Matter

Unit Experiment		**E1**

CHAPTER 1 — **Matter and Its Properties** …………………………… **E2**

Lesson 1—How Can Physical Properties Be Used to Identify Matter? ……E4
Lesson 2—How Does Matter Change from One State to Another? ……E12
Lesson 3—How Does Matter React Chemically? …………………………E20
  Science and Technology • Self-Healing Asphalt …………………………E28
  People in Science • Theophilus Leapheart …………………………………E30
  Activities for Home or School ………………………………………………E31
Chapter Review and Test Preparation …………………………………………E32

CHAPTER 2 — **Atoms and Elements** …………………………………… **E34**

Lesson 1—What Are Atoms and Elements? ………………………………E36
Lesson 2—What Are Compounds? …………………………………………E44
  Science Through Time • Discovering Elements …………………………E50
  People in Science • Lise Meitner ……………………………………………E52
  Activities for Home or School ………………………………………………E53
Chapter Review and Test Preparation …………………………………………E54

Unit Expeditions …………………………………………………………… **E56**

UNIT F — PHYSICAL SCIENCE
Energy and Motion

Unit Experiment	**F1**

CHAPTER 1 — Forces — F2
Lesson 1—What Forces Affect Objects on Earth Every Day? F4
Lesson 2—What Are Balanced and Unbalanced Forces? F10
Lesson 3—What Is Work and How Is It Measured? F16
    Science and Technology • Weightless Work F24
    People in Science • Ephraim Fishbach F26
    Activities for Home or School F27
Chapter Review and Test Preparation F28

CHAPTER 2 — Motion — F30
Lesson 1—How Are Motion and Speed Related? F32
Lesson 2—What Are the Three Laws of Motion? F38
Lesson 3—Why Do the Planets Stay in Orbit? F46
    Science and Technology • Skipping Through Space F52
    People in Science • Patricia Cowings F54
    Activities for Home or School F55
Chapter Review and Test Preparation F56

CHAPTER 3 — Forms of Energy — F58
Lesson 1—What Are Kinetic and Potential Energy? F60
Lesson 2—What Is Electric Energy? F66
Lesson 3—What Are Light and Sound Energy? F74
Lesson 4—What Are Thermal and Chemical Energy? F82
    Science Through Time • Sounds and Images F88
    People in Science • Jean M. Bennett F90
    Activities for Home or School F91
Chapter Review and Test Preparation F92

CHAPTER 4 — How People Use Energy — F94
Lesson 1—How Do People Use Fossil Fuels? F96
Lesson 2—How Can Moving Water Generate Electricity? F102
Lesson 3—What Other Sources of Energy Do People Use? F108
    Science and Technology • Canola Motor Oil F114
    People in Science • Meredith Gourdine F116
    Activities for Home or School F117
Chapter Review and Test Preparation F118

Unit Expeditions	**F120**

HOW SCIENTISTS WORK

Planning an Investigation

How do scientists answer a question or solve a problem they have identified? They use organized ways called **scientific methods** to plan and conduct a study. They use science process skills to help them gather, organize, analyze, and present their information.

Justin is using this scientific method for experimenting to find an answer to his question. You can use these steps, too.

STEP 1 Observe and ask questions.

- Use your senses to make observations.
- Record **one** question that you would like to answer.
- Write down what you already know about the topic of your question.
- Decide what other information you need.
- Do research to find more information about your topic.

What design of paper airplane will fly the greatest distance? I need to find out more about airplane wings.

STEP 2 Form a hypothesis.

- Write a possible answer to your question. A possible answer to a question that can be tested is a **hypothesis**.
- Write your hypothesis in a complete sentence.

My hypothesis is: This airplane, with the narrow wings, will fly farthest.

STEP 3 Plan an experiment.

- Decide how to conduct a fair test of your hypothesis by controlling variables. **Variables** are factors that can affect the outcome of the investigation.
- Write down the steps you will follow to do your test.
- List the equipment you will need.
- Decide how you will gather and record your data.

I'll launch each airplane three times. Each airplane will be launched from the same spot, and I'll use the same amount of force each time.

STEP 4 Conduct the experiment.

- Follow the steps you wrote.
- Observe and measure carefully.
- Record everything that happens.
- Organize your data so you can study it carefully.

I'll record each distance. Then I'll find the average distance each airplane traveled.

HOW SCIENTISTS WORK

STEP 5 Draw conclusions and communicate results.

- Analyze the data you gathered.
- Make charts, tables, or graphs to show your data.
- Write a conclusion. Describe the evidence you used to determine whether your test supported your hypothesis.
- Decide whether your hypothesis was correct.

My hypothesis was correct. The airplane with the narrow wings flew farthest.

INVESTIGATE FURTHER

What if your hypothesis was correct . . .

You may want to pose another question about your topic that you can test.

What if your hypothesis was incorrect . . .

You may want to form another hypothesis and do a test on a different variable.

I'll test this new hypothesis: The airplane with the narrow wings will also fly for the longest time.

Do you think Justin's new hypothesis will be correct? Plan and conduct a test to find out!

Using Science Process Skills

When scientists try to find an answer to a question or do an experiment, they use thinking tools called **process skills**. You use many of the process skills whenever you speak, listen, read, write, or think. Think about how these students use process skills to help them answer questions, do experiments, and investigate the world around them.

HOW SCIENTISTS WORK

What Greg plans to investigate

Greg is finding leaves in the park. He wants to make collections of leaves that are alike in some way. He looks for leaves of different sizes and shapes.

Process Skills

Observe—use the senses to learn about objects and events

Compare—identify characteristics about things or events to find out how they are alike and different

Measure—compare an attribute of an object, such as its mass, length, or volume, to a standard unit, such as a gram, centimeter, or liter

Classify—group or organize objects or events in categories based on specific characteristics

How Greg uses process skills

He **observes** the leaves and **compares** their sizes, shapes, and colors. He **measures** each leaf with a ruler. Then he **classifies** the leaves, first into groups based on their sizes and then into groups based on their shapes.

What Pilar plans to investigate

It's been raining for part of the week. Pilar wants to know if it will rain during the coming weekend.

How Pilar uses process skills

She **gathers and records data** to make a prediction about the weather. She observes the weather each day of the week and records it. On a chart, she **displays data** she has gathered. On Friday, she **predicts**, based on her observations, that it will rain during the weekend.

Process Skills

Gather, Record, Display, or Interpret Data

- gather data by making observations and use them to make inferences or predictions
- record data by writing down observations
- display data by making tables, charts, or graphs

Predict—form an idea of an expected outcome, based on observations or experience

HOW SCIENTISTS WORK

Process Skills

Hypothesize—make a statement about an expected outcome, based on observation, knowledge, and experience

Plan and Conduct a Simple Investigation—identify and perform the steps necessary to find the answer to a question, using appropriate tools and recording and analyzing the data collected

Infer—use logical reasoning to explain events and draw conclusions based on observations

What Tran plans to investigate

Tran is interested in knowing how the size of a magnet is related to its strength.

How Tran uses process skills

He **hypothesizes** that larger magnets are stronger than smaller magnets. He **plans and conducts a simple investigation** to see if his hypothesis is correct. He gathers magnets of different sizes and objects of different weights that the magnets will attract. Tran tests each item with each magnet and records his findings. His hypothesis seems to be correct until he tests the last object, a toy truck. When the large bar magnet cannot pick up the truck, but the smaller horseshoe magnet can, he **infers** that the largest magnets are not always the strongest.

What Emily plans to investigate

Emily sees an ad about food wrap. The people in the ad claim that Tight-Right food wrap seals containers better than other food wraps. Emily plans a simple experiment to find out if this claim is true.

How Emily uses process skills

She **identifies and controls variables** in the experiment by choosing three bowls that are exactly the same. She labels the bowls A, B, and C, places them on a tray, and adds exactly 350 mL of water to each bowl. She cuts a 25-cm-long piece of Tight-Right food wrap and covers bowl A. She cuts 25-cm-long pieces of two other brands of food wrap and covers bowls B and C. She seals the food wrap on all three bowls as tightly as she can. Emily **experiments** with the seals by shaking the tray on which the bowls are placed. Water sloshes up the sides of the bowls and leaks out onto the tray from bowls B and C, but not from bowl A. From her observations she infers that the claim for Tight-Right food wrap is true.

Process Skills

Identify and Control Variables—identify and control factors that affect the outcome of an experiment

Experiment—design ways to collect data to test hypotheses under controlled conditions

HOW SCIENTISTS WORK

Reading to Learn

Scientists use reading, writing, and numbers in their work. They **read** to find out everything they can about a topic they are investigating. So it is important that scientists know the meanings of science vocabulary and that they understand what they read. Use the following strategies to help you become a good science reader!

Before Reading

- Read the **Find Out** statements to help you know what to look for as you read.
- Think: I need to find out how living things get the energy they need.

- Look at the **Vocabulary** terms.
- Be sure you can pronounce each term.
- Look up each term in the Glossary.
- Say the definition to yourself. Use the term in a sentence to show its meaning.

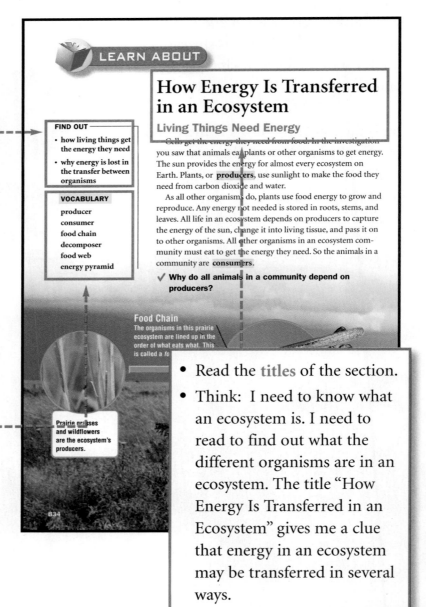

- Read the **titles** of the section.
- Think: I need to know what an ecosystem is. I need to read to find out what the different organisms are in an ecosystem. The title "How Energy Is Transferred in an Ecosystem" gives me a clue that energy in an ecosystem may be transferred in several ways.

xviii

During Reading

Find the **main idea** in the first paragraph.

- In food chains, there are more producers than consumers.

Find the **details** in the next paragraph that support the main idea.

- Only 10 percent of the energy at any level is passed on to the next level.
- High-level consumers, such as wolves, have relatively small populations. There isn't enough energy for large populations of wolves.
- All other organisms in an ecosystem must eat to get the energy they need.

Energy Pyramids

In the food chains of most ecosystems, there are many more producers than there are consumers. Producers use about 90 percent of the food energy they make during photosynthesis for their life processes. Only 10 percent of the energy is stored in plant tissue. When a consumer eats the plant tissue, it uses about 90 percent of the plant's stored food energy to stay alive. It stores the other 10 percent in its body tissue. This huge loss of stored food energy occurs at each level in a food chain. An **energy pyramid** shows the amount of energy available to pass from one level of a food chain to the next.

Remember, only 10 percent of the energy at any level of a food chain is passed on to the next higher level. Since less energy is available to organisms higher up the food chain, there are usually fewer organisms at these levels. High-level consumers, such as wolves, have relatively small populations. There is not enough energy available to support a large population of wolves.

The size of each level of an energy pyramid is related to the sizes of the populations at that level. The producer population is usually the largest, since it provides energy for all consumer levels in the pyramid.

✓ How much of the food energy that is taken in by an organism is used for its own life processes?

Energy Pyramid

◄ **Third-level consumers** Hawks are at the top of this energy pyramid. They eat snakes. There are few hawks because most of the energy has been used at lower levels of the pyramid.

◄ **Second-level consumers** There are far fewer snakes than grasshoppers. This is because grasshoppers use 90 percent of the food energy they get for their own life processes.

◄ **First-level consumers** Since plants use 90 percent of the food energy they produce, there are fewer grasshoppers than there are grasses and other plants.

◄ **Producers** Producers, such as grasses and other plants, form the base of an energy pyramid.

Check your understanding of what you have read.

- Answer the question at the end of the section.
- If you're not sure of the answer, reread the section and look for the answer to the question.

HOW SCIENTISTS WORK

After Reading

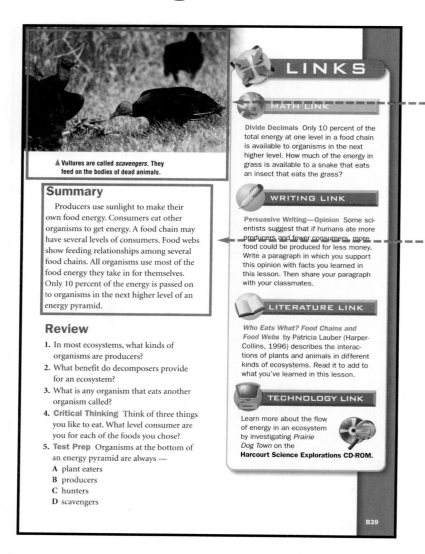

Study the photographs.
- Read the caption with each photograph.
- Think: What kinds of organisms are shown in the photographs? What do they eat?

Summarize what you have read.
- Think about what you've already learned about energy transfer in an ecosystem.
- Ask yourself: Why does an ecosystem depend on its producers? What kinds of foods do the consumers eat?

For more reading strategies and tips, see pages R38–R49.

Reading about science helps you understand your conclusions from your investigations.

Writing to Communicate

Writing about what you are learning helps you connect the new ideas to what you already know. Scientists **write** about what they learn in their research and investigations to help others understand the work they have done. As you work like a scientist, you will use the following kinds of writing to describe what you are doing and learning.

In **informative writing**, you may
- describe your observations, inferences, and conclusions.
- tell how to do an experiment.

In **narrative writing**, you may
- describe something, give examples, or tell a story.

In **expressive writing**, you may
- write letters, poems, or songs.

In **persuasive writing**, you may
- write letters about important issues in science.
- write essays expressing your opinions about science issues.

Writing about what you have learned about science helps others understand your thinking.

HOW SCIENTISTS WORK

Using Numbers

Scientists **use numbers** when they collect, display, and interpret their data. Understanding numbers and using them correctly to show the results of investigations are important skills that a scientist must have. As you work like a scientist, you will use numbers in the following ways:

Measuring

Scientists make accurate measurements as they gather data. They use measuring instruments such as thermometers, clocks and timers, rulers, a spring scale, and a balance, and they use beakers and other containers to measure liquids.

For more information about using measuring tools, see pages R2–R6.

Interpreting Data

Scientists collect, organize, display, and interpret data as they do investigations. Scientists choose a way to display data that helps others understand what they have learned. Tables, charts, and graphs are good ways to display data so that it can be interpreted by others.

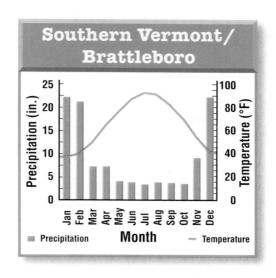

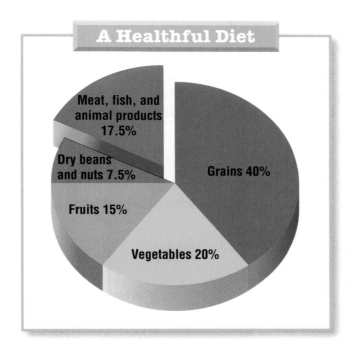

Using Number Sense

Scientists must understand what the numbers they use represent. They compare and order numbers, compute with numbers, read and understand the numbers shown on graphs, and read the scales on thermometers, measuring cups, beakers, and other tools.

Good scientists apply their math skills to help them display and interpret the data they collect.

In *Harcourt Science* you will have many opportunities to work like a scientist. An exciting year of discovery lies ahead!

HOW SCIENTISTS WORK

Safety in Science

Doing investigations in science can be fun, but you need to be sure you do them safely. Here are some rules to follow.

1. **Think ahead.** Study the steps of the investigation so you know what to expect. If you have any questions, ask your teacher. Be sure you understand any safety symbols that are shown.

2. **Be neat.** Keep your work area clean. If you have long hair, pull it back so it doesn't get in the way. Roll or push up long sleeves to keep them away from your experiment.

3. **Oops!** If you should spill or break something, or get cut, tell your teacher right away.

4. **Watch your eyes.** Wear safety goggles anytime you are directed to do so. If you get anything in your eyes, tell your teacher right away.

5. **Yuck!** Never eat or drink anything during a science activity.

6. **Don't get shocked.** Be especially careful if an electric appliance is used. Be sure that electric cords are in a safe place where you can't trip over them. Don't ever pull a plug out of an outlet by pulling on the cord.

7. **Keep it clean.** Always clean up when you have finished. Put everything away and wipe your work area. Wash your hands.

In some activities you will see these symbols. They are signs for what you need to be safe.

Be especially careful.

Wear safety goggles.

Be careful with sharp objects.

Don't get burned.

Protect your clothes.

Protect your hands with mitts.

Be careful with electricity.

UNIT A — LIFE SCIENCE

Living Systems

CHAPTER 1   From Single Cells to Body Systems....A2

CHAPTER 2   Classifying Living Things A34

CHAPTER 3   Animal Growth and Heredity A60

CHAPTER 4   Plants and Their Adaptations...........A88

Unit Expeditions.........................A120

UNIT EXPERIMENT

Plants and Light

Living organisms respond to certain factors in their environments. One environmental factor that plants respond to is light. While you study this unit, you can conduct a long-term experiment related to this response. Here are some questions to think about. How do plants respond to light? For example, will plants grow toward a light source? Plan and conduct an experiment to find answers to these or other questions you have about plants and light. See pages x–xvii for help in designing your experiment.

CHAPTER 1
From Single Cells to Body Systems

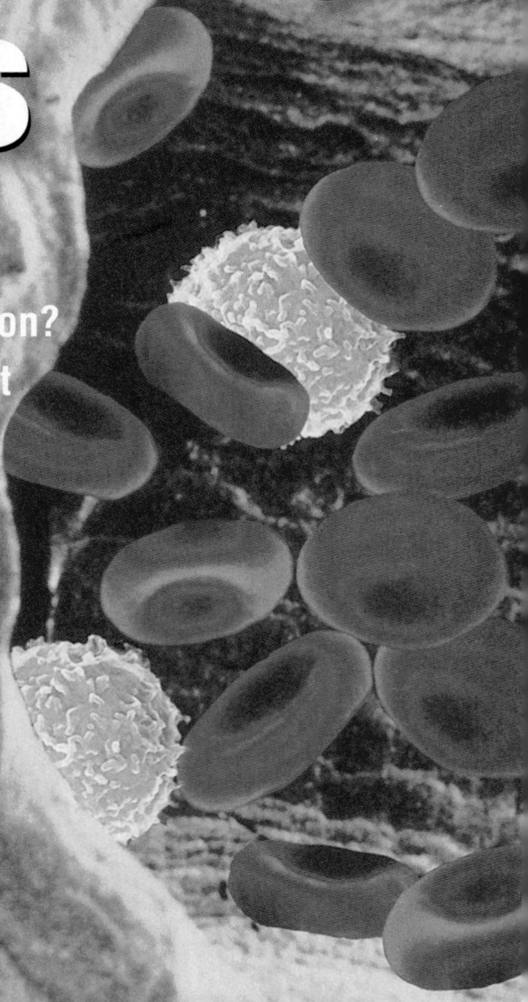

Vocabulary Preview

cell
cell membrane
nucleus
cytoplasm
diffusion
osmosis
tissue
organ
system
capillaries
alveoli
villi
nephrons
bone marrow
joints
tendons
ligaments
neuron
receptors

Do you know what a fish, a tree, and a human being have in common? They are all made of tiny cells that carry on the processes of life.

Fast Fact

A human body contains about 70,000 miles of blood vessels. Within the blood vessel shown here are disk-shaped red blood cells and round white blood cells.

Fast Fact

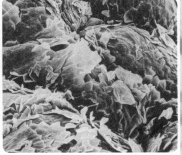

A scanning electron microscope (SEM) can magnify objects as much as 900,000 times. This electron micrograph shows human face cells magnified 100,000 times.

Fast Fact

A single square inch of human skin has more than 19 million cells.

LESSON 1

What Are Cells, and What Do They Do?

In this lesson, you can . . .

 INVESTIGATE what cells look like.

 LEARN ABOUT cells.

 LINK to math, writing, health, and technology.

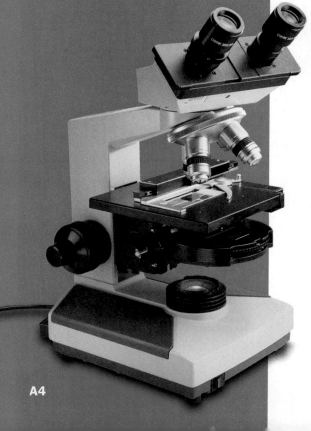

Observing Cells

Activity Purpose If you're looking at a landscape from far away, you might use a telescope to make the details clearer. Suppose you focus the telescope on a distant farm. You can see crates of freshly harvested onions. Now suppose you use a microscope to magnify the scene more and more. What details about an onion might you **observe**? In this investigation you'll observe a thin layer of an onion skin. Then you will observe and **compare** other plant cells and animal cells.

Materials
- Microslide Viewer
- Microslide of cell structure
- colored pencils

Alternate Materials
- slice of onion
- microscope slide
- coverslip
- dropper
- red food coloring
- microscope
- colored pencils

Activity Procedure

1. Insert the Cell Structure Microslide in the slot on the Microslide Viewer. Turn the focus knob until you can see the cells clearly. (Picture A)

2. **Observe** the onion skin cells and the human cheek cells. **Record** your observations by using the colored pencils to make drawings.

◀ This microscope allows a person to study a thin slice of material under high magnification.

3. Now **observe** the green leaf cells and the nerve cells. Again, **record** your observations by making drawings. (Picture B)

4. Now **compare** your drawings. Make a Venn diagram with two large, overlapping circles. Label the circles *Plant Cells* and *Animal Cells*. Label the area where the circles overlap *Both Cells*. Draw the cell parts that you **observed** in the proper circles. Leave enough room to label the parts as you read about them in this lesson.

Picture A

Draw Conclusions

1. **Compare** the outer layers of plant and animal cells.

2. In the centers of most cells are structures that determine how cells function. How many of these structures are there in each of the cells you **observed**?

3. **Scientists at Work** Scientists often **infer** characteristics of a group of objects by **observing** just a few of the objects. From your observations, what do you infer about the number of determining structures in a cell?

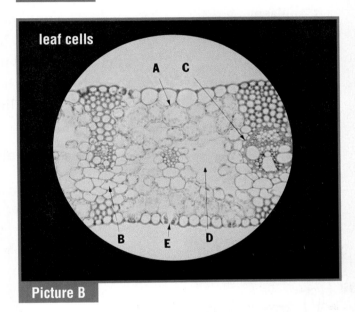

Picture B

Investigate Further Now that you have **observed** photomicrographs of cells, what questions do you have about living cells? Use the materials in the *Alternate Materials* list to **plan and conduct a simple experiment** based on this hypothesis: All cells have certain parts in common. See page R3 for tips on using a microscope.

Process Skill Tip

You can **infer** based on what you **observe**, or based on other information you have about a subject.

LEARN ABOUT

Cells

FIND OUT
- what cells are
- how cells are organized
- what cells do

VOCABULARY

cell
cell membrane
nucleus
cytoplasm
diffusion
osmosis
tissue
organ
system

The Discovery of Cells

The Microslide and viewer you used in the investigation enabled you to observe parts of plants and animals under magnification. Without magnification, you couldn't have seen the structures you did. A microscope magnifies objects in a similar way. In fact, the photomicrographs you observed were taken through a microscope. The first microscopes were invented in the early 1600s. One scientist who built and used an early microscope was Robert Hooke.

In 1665 Hooke observed a thin slice of cork through a microscope. The tiny walled spaces he saw in the cork reminded him of tiny rooms. So he called them cells. Over the next 200 years, scientists learned more and more about cells. They learned that the **cell** is the basic unit of structure and function of all living things. The time line below shows some important early discoveries about cells.

✓ Why were cells not observed before the 1600s?

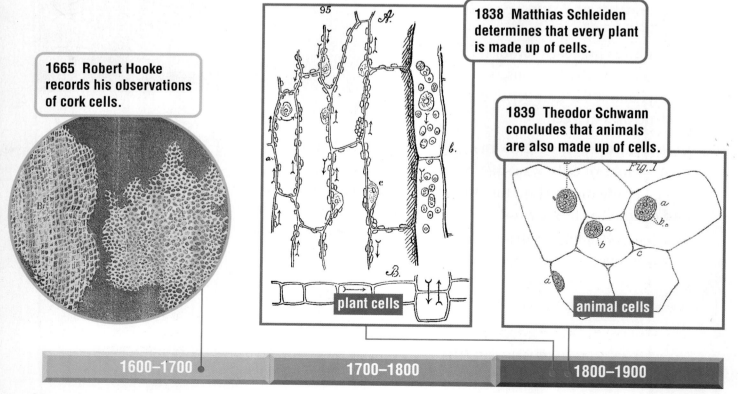

1665 Robert Hooke records his observations of cork cells.

1838 Matthias Schleiden determines that every plant is made up of cells.

1839 Theodor Schwann concludes that animals are also made up of cells.

plant cells

animal cells

1600–1700 | 1700–1800 | 1800–1900

Kinds of Cells

Scientists have classified about a million kinds of plants and animals. But as different as those plants and animals seem to be, all of them are made of cells.

The simplest organisms, such as bacteria, are each a single cell. Most plants and animals, however, are made up of many cells. Humans, for example, are made up of *trillions* of cells. An organism with many cells usually has many different kinds of cells. Each kind of cell has a particular function.

The size and shape of a cell depend on its function. Red blood cells, for example, are small and disc-shaped. They can easily fit through the smallest blood vessels. Muscle cells are long and thin. When they contract, or shorten, they produce movement. Nerve cells, which carry signals from the brain to the muscles, are very long.

Plants also have different kinds of cells. Some plant cells take in water from the soil. Others protect the plant. And still others make food.

Cells work together to perform basic life processes that keep an organism alive. These processes include releasing energy from food, getting rid of body wastes, and making new cells for growth and repair. In addition to performing a particular function for the organism, each individual cell can perform all the basic life processes for itself.

✓ **Why might bone cells be different from muscle cells?**

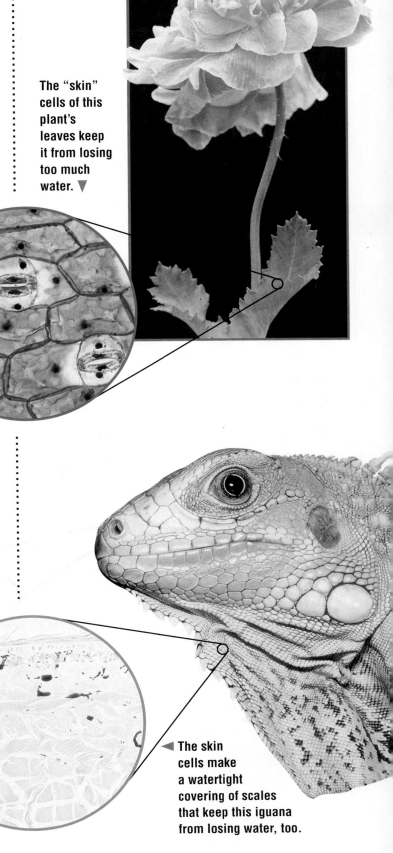

The "skin" cells of this plant's leaves keep it from losing too much water. ▼

◀ The skin cells make a watertight covering of scales that keep this iguana from losing water, too.

Plant and Animal Cells

Although cells are the basic unit of all living organisms, cells contain even smaller structures called *organelles* (awr•guh•NELZ). Each organelle has a particular function in the life processes of a cell.

All cells—except those of bacteria—have similar organelles. For example, every cell is enclosed by a thin covering called the **cell membrane**. The cell membrane holds the parts of the cell together. It also separates the cell from its surroundings.

Most cells have a nucleus (NOO•klee•uhs). The **nucleus**, which is enclosed in its own membrane, determines the cell's activities. One function of the nucleus is to control cell reproduction. Cells can grow only to a certain size. So the number of cells has to increase in order for plants and animals to grow.

Inside the nucleus are threadlike structures called *chromosomes* (KROH•muh•sohmz).

THE INSIDE STORY

Comparing Plant and Animal Cells

Plant cells have different shapes and sizes, but they all have the same parts. The diagram shows what you might observe if you could look inside a leaf cell. The organelles you see are working parts of a complete cell. Each organelle has its own specific function.

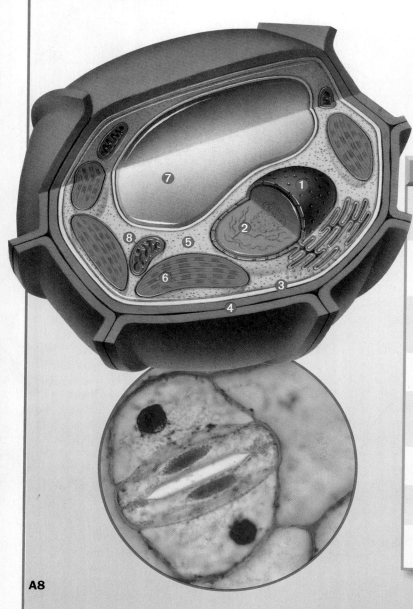

Plant Cell Structures

Nucleus—①	the organelle that determines all of a plant cell's activities and the production of new cells
Chromosomes—②	threadlike structures that contain information about the characteristics of the plant
Cell membrane—③	a covering that holds the plant cell together and separates it from its surroundings
Cell wall—④	a rigid layer that supports and protects the plant cell
Cytoplasm—⑤	a jellylike substance that contains many chemicals to keep the cell functioning
Chloroplasts—⑥	organelles that make food for the plant cell
Vacuole—⑦	an organelle that stores food, water, or wastes
Mitochondria—⑧	organelles that release energy from food

These contain information about the characteristics of the organism. When a cell reproduces, identical chromosomes go into each new cell.

Between the cell membrane and the nucleus is the cytoplasm (SYT•oh•plaz•uhm). **Cytoplasm** is a jellylike substance containing many chemicals to keep the cell functioning.

There are several kinds of organelles in the cytoplasm. Each is enclosed in a membrane. *Mitochondria* (myt•oh•KAHN•dree•uh) release energy from food. *Vacuoles* (VAK•yoo•ohlz) are storage organelles. They store food, water, or waste materials.

Two organelles make plant cells different from animal cells. In addition to a cell membrane, a plant cell is surrounded by a rigid *cell wall*, which gives it strength. Plant cells also have *chloroplasts*, which make food.

✓ **How do plant cells and animal cells differ?**

The functions that allow an animal to live and grow are also carried out in its cells. Although the iguana's skin cells are a different shape and size from its blood cells, each cell typically contains the same parts. As you look at the diagram of an animal cell, notice how it differs from the plant cell.

Animal Cell Structures

Nucleus— ❶	the organelle that determines all of an animal cell's activities and the production of new cells
Chromosomes— ❷	threadlike structures that contain information about the characteristics of the animal
Cell membrane— ❸	a covering that holds the animal cell together and separates it from its surroundings
Cytoplasm— ❹	a jellylike substance that contains many chemicals to keep the cell functioning
Vacuoles— ❺	organelles that store food, wastes, or water
Mitochondria— ❻	organelles that release energy from food

Materials Move into and out of Cells

Most of the activities of a cell require energy. That energy is supplied by the mitochondria. Mitochondria use food, oxygen, and water to produce energy. This process produces carbon dioxide. How do cells get needed materials like food, water, and oxygen? And how do they get rid of wastes, like carbon dioxide?

Many materials move into and out of cells by diffusion. In the process of **diffusion**, particles of a substance move from an area where there are a lot of particles of the substance to an area where there are fewer particles of the substance.

For example, red blood cells carry oxygen from the lungs to all parts of the body. There is a lot of oxygen in red blood cells and very little in other body cells. So oxygen diffuses out of red blood cells and into body cells. At the same time, there is a lot of carbon dioxide in body cells and very little in the blood. So carbon dioxide diffuses out of body cells and into the blood.

Diffusion of materials into and out of cells takes place through the cell membrane. You might think of a cell membrane as a filter. It allows some particles to pass through, but it keeps other particles out. Water and materials dissolved in water—such as sugar—diffuse easily through cell membranes. Diffusion doesn't require energy from the cell.

The movement of water and dissolved materials through cell membranes is so important to living organisms that it is given a special name— **osmosis**. Cells get most of their water by osmosis.

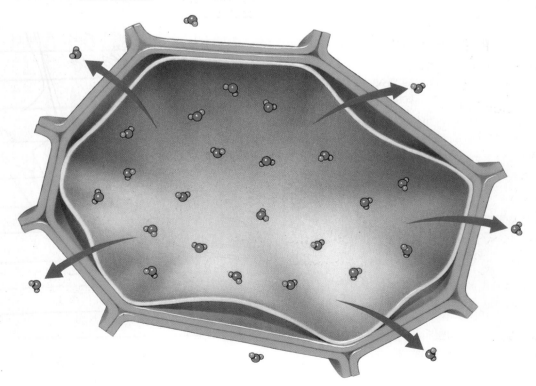

When a plant gets too little water, water leaves the cells because the concentration of water is higher inside the cells than outside.

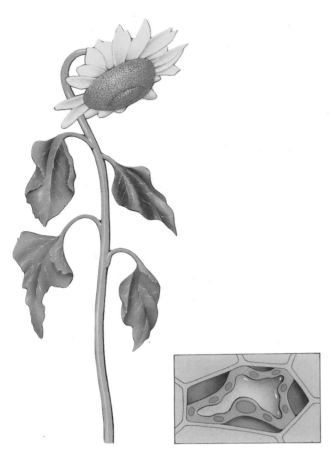

▲ If the soil is dry, water moves from the plant's cells into the soil. The cells shrink, and the plant wilts.

▲ If the soil is moist, water moves back into the plant's cells. The cells expand, and the plant recovers.

By drinking, animals replace the water their cells lose. ▼

Osmosis keeps plants from wilting. Since there is usually more water in the soil than in the roots of plants, water flows into plant cells. Water flowing into plant cells fills the vacuoles, which pushes the cytoplasm tightly against the cell walls. This causes plant stems and leaves to stand straight.

If the soil is very dry, there is more water in the plant than in the soil. Water leaves the plant by osmosis. The loss of water from the plant's vacuoles causes cytoplasm to shrink away from cell walls. The plant wilts and may die if it loses too much water.

✓ **What is osmosis?**

Tissues, Organs, and Systems

In an organism made up of many cells, similar cells work together. Cells that work together to perform a specific function form a **tissue**. There are four kinds of tissues in humans.

Most of the mass of an animal is *muscle tissue*. Muscle tissue is made up of cells that contract when they receive signals from the brain. The contraction and relaxation of muscle tissue move the skeleton. The signals that cause muscle tissue to contract travel through another kind of tissue—*nervous tissue*. The brain and spinal cord, as well as the places where sight, hearing, taste, smell, and touch begin, are all nervous tissue.

Connective tissue is the third kind of tissue. It includes the tissue in bones, cartilage, and tendons. Blood is also a connective tissue.

The final kind of tissue is *epithelial* (ep•ih•THEE•lee•uhl) *tissue*. Epithelial tissue includes the body covering of an animal. It also lines most internal organs.

Just as cells that work together form a tissue, tissues that work together form an **organ**. Each organ in an animal's body is made of several kinds of tissues. Skin, for example, is an organ. It is made of many layers of epithelial tissue, as well as muscle tissue, nervous tissue, and a cushioning layer of connective tissue.

Each organ in an animal's body performs a major function that keeps the animal alive. The heart, for example, is an organ that pumps blood throughout the animal's body.

Organs that work together to perform a function form a **system**. A human has ten

In this diagram, you can see the four levels of organization in the digestive system. ▼

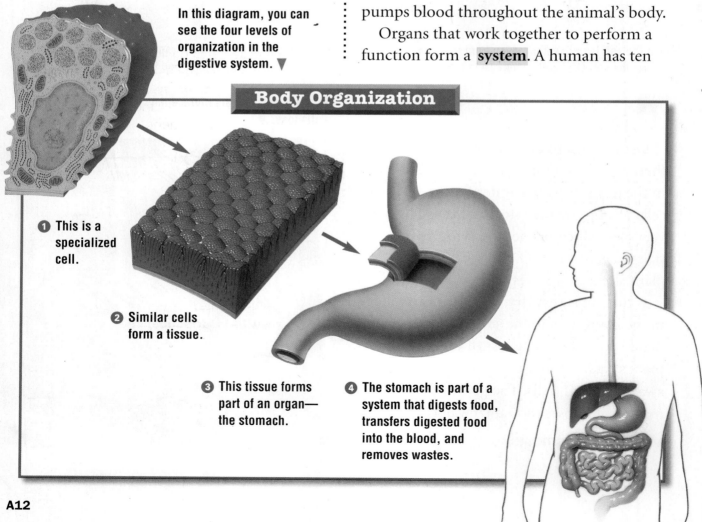

Body Organization

❶ This is a specialized cell.

❷ Similar cells form a tissue.

❸ This tissue forms part of an organ—the stomach.

❹ The stomach is part of a system that digests food, transfers digested food into the blood, and removes wastes.

major body systems. You will learn about some of those systems in the next two lessons.

Plant cells also form tissues, such as the bark of a tree. And plant tissues work together, forming organs, such as roots and leaves. You will learn more about plants in Chapters 2 and 4.

✓ **What kind of tissue gives an organism the ability to move?**

Summary

All living things are made up of one or more cells. Each cell is able to perform the functions that support life. Plant cells differ from animal cells in that they have cell walls and chloroplasts. Cells obtain the materials they need through the cell membrane. Cells with similar functions form tissues. Tissues that function together make up an organ. Organs working together form a body system.

Review

1. What is the function of a cell's nucleus?
2. How do vacuoles in plant cells help keep the plant upright?
3. What is the difference between diffusion and osmosis?
4. **Critical Thinking** If you stand at one end of a room and spray perfume into the air, a person at the other end of the room will soon smell the perfume. Explain.
5. **Test Prep** One example of a tissue is —
   A heart   C muscle
   B nucleus   D mitochondria

LINKS

MATH LINK

Solve Problems Suppose a single cell divides into two cells every 15 min. If each of those also divides into two, and so on, how long will it take for a single cell to produce 500 cells?

WRITING LINK

Informative Writing—Compare and Contrast Write two or three paragraphs for your teacher comparing plant cells and animal cells.

HEALTH LINK

Interview a Doctor Find out how studying cells helps doctors understand the human body. Brainstorm with a family member to come up with questions you can ask your doctor the next time you have an appointment.

TECHNOLOGY LINK

Learn more about cells and body systems by using the activities and information provided on the Harcourt Learning Site.
www.harcourtschool.com

LESSON 2

How Do Body Systems Transport Materials?

In this lesson, you can . . .

 INVESTIGATE cells and tissues.

 LEARN ABOUT four human body systems.

 LINK to math, writing, social studies, and technology.

◀ The materials your body uses to produce energy must be replaced.

INVESTIGATE

Cells and Tissues

Activity Purpose Your body is made up of cells that are organized into tissues, organs, and systems. Cells are highly specialized for the functions they perform for the body. Even cells that make up similar tissues can differ in many ways. In this investigation you will **observe** and **compare** under magnification several kinds of cells and tissues.

Materials
- Microslide Viewer
- Microslide of animal tissues
- colored pencils

Alternate Materials
- prepared slides of epithelial, connective, and nervous tissues
- microscope

Activity Procedure

1. Insert the Animal Tissues Microslide in the slot of the Microslide Viewer. Turn the focus knob until you can see the cells and tissues clearly. (Picture A)

2. **Observe** the voluntary muscle cells. **Record** your observations by using the colored pencils to make a drawing. Label your drawing with the name of the tissue. Then describe the tissue. You may use the Microslide text folder to help you write your description. (Picture B)

Picture A

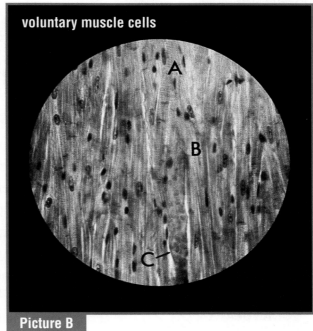
Picture B

3. Repeat Step 2 for the smooth muscle cells and the heart muscle.

4. **Compare** the three kinds of muscle tissue.

Draw Conclusions

1. How are the three kinds of muscle tissue alike? How are they different?

2. The dark-stained organelles you **observed** in the muscle tissues are mitochondria. Which kind of muscle tissue has the most mitochondria?

3. **Scientists at Work** When scientists **compare** objects, they often **infer** reasons for any differences. What do you infer about why one kind of muscle tissue has more mitochondria than the others?

Process Skill Tip

Making drawings of objects allows you to **compare** them and **infer** reasons for differences between them.

Investigate Further Now that you have **observed** several kinds of tissues, develop a testable question about differences among tissues. Use the materials in the *Alternate Materials* list to study other kinds of tissues. Observe the tissues under the microscope, and draw and label any differences you see. **Form a hypothesis** about how these tissues are different from the muscle tissues you observed. See page R3 for tips on using a microscope.

LEARN ABOUT

FIND OUT

- about the circulatory, respiratory, digestive, and excretory systems
- which organs make up each system
- how the systems work together

VOCABULARY

capillaries
alveoli
villi
nephrons

Human Body Systems

From Cells to Systems

The tissues you observed in the investigation combine in various ways to form body organs. Certain organs work together to form body systems. Each body system has a specific task that helps keep you alive. But your body systems also work together. On its own, a single body system cannot keep you alive.

The digestive system, for example, breaks down food into nutrients the body needs for energy. But without the circulatory system, the nutrients couldn't travel to the parts of the body that need them. The circulatory system also delivers the oxygen needed to release energy from food. However, without the respiratory system, oxygen couldn't get into the circulatory system. And all body processes produce wastes that must be removed. The excretory and respiratory systems share this function, with transportation provided by the circulatory system.

As you can see, working together is important for living organisms. Cells work together to form tissues. Tissues work together to form organs. Organs work together to form systems. And systems work together to keep you alive.

✓ **How do body systems depend on each other?**

These red blood cells are part of a tissue—blood. ▼

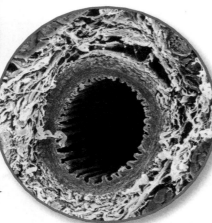

Blood vessels are one kind of organ in the circulatory system. ▶

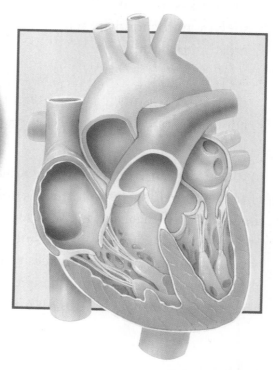

The heart is another organ of the circulatory system. It is made of muscle tissue. ▶

The Circulatory System

The circulatory system transports oxygen, nutrients, and wastes through the body in the blood. The liquid part of blood, called *plasma*, is mostly water. Plasma also contains dissolved nutrients and waste products such as carbon dioxide. The solid part of blood includes red blood cells and white blood cells. Red blood cells absorb oxygen from air in the lungs and transport it to every cell in the body. White blood cells help the body fight infection. They attack and destroy viruses and bacteria that enter the body.

Blood also contains *platelets*—tiny pieces of blood cells inside membranes. Platelets cause blood to clot when a blood vessel is cut. They also help repair damage to blood vessels.

The heart, an organ made of muscle tissue, pumps blood through blood vessels. The heart has four chambers, or parts. Oxygen-rich blood from the lungs enters one chamber. It moves to the next chamber, from which it is pumped to the body. Oxygen-poor blood from the body enters the third chamber. It moves to the fourth chamber, from which it is pumped to the lungs.

Blood leaves the heart through blood vessels called *arteries*. Arteries lead to capillaries. **Capillaries** are blood vessels so small that blood cells have to move through them in single file. There are capillaries throughout the body, so nutrients and oxygen can reach every cell. Waste products from cells are picked up by plasma in the capillaries. Capillaries lead to larger vessels, called *veins*, which return blood to the heart.

✓ **Why are platelets important?**

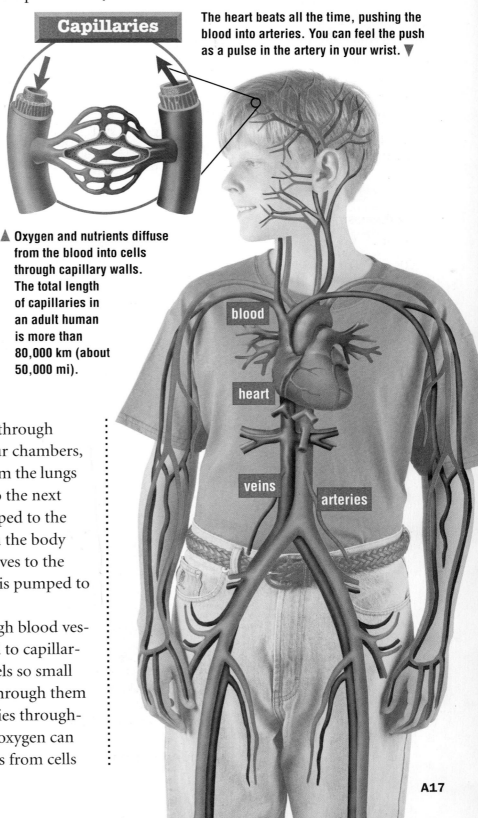

▲ Oxygen and nutrients diffuse from the blood into cells through capillary walls. The total length of capillaries in an adult human is more than 80,000 km (about 50,000 mi).

The heart beats all the time, pushing the blood into arteries. You can feel the push as a pulse in the artery in your wrist. ▼

The Respiratory System

Your body uses a lot of energy. So your cells need a lot of food and oxygen. You get the oxygen your cells need by breathing. When you inhale, several liters of air are pulled into your body. The air is filtered by tiny hairs in your nose and warmed by capillaries that line the nasal passages. Warm, clean air then travels down your *trachea*, or windpipe.

The lungs are the major organs of the respiratory system. ▼

trachea
bronchi

In your chest, the trachea branches into two tubes called *bronchi* (BRAHNG•kee). Each tube leads into a lung. In the lungs, the bronchi divide into smaller and smaller tubes. At the end of the smallest tubes are tiny air sacs called **alveoli** (al•VEE•oh•lee). The walls of the alveoli are only one cell thick and are surrounded by capillaries.

The capillaries surrounding the alveoli get blood from the *pulmonary arteries* coming from the heart. This blood contains a lot of carbon dioxide. Carbon dioxide is a waste produced by the process that releases energy in cells. Carbon dioxide diffuses through the thin walls of the alveoli and into air that will be exhaled. At the same time, oxygen from inhaled air diffuses through the alveoli and into red blood cells in the capillaries. The oxygen-rich blood then flows from the capillaries into the *pulmonary veins* and back to the heart. From the heart, oxygen-rich blood is pumped to other parts of the body.

✓ **What happens in the alveoli?**

▲ The gas exchange of oxygen and carbon dioxide takes place in the alveoli.

The lungs are on a separate circuit of the circulatory system. The heart pumps oxygen-poor blood through the pulmonary arteries to the lungs. The oxygen-rich blood travels through the pulmonary veins back to the heart. ▼

The Digestive System

Your digestive system provides the nutrients your cells need to produce energy. To provide nutrients, the digestive system performs two functions. The first is to break food into nutrients. The second is to get the nutrients into the blood. Then the circulatory system transports them to your cells.

Digestion begins as you chew food, breaking it into smaller pieces so that you can swallow it. Glands in your mouth produce saliva. Saliva moistens food and begins to break down starchy foods, such as pasta, into sugars. (If you chew an unsalted cracker for a while, it will begin to taste sweet.)

When you swallow, food passes through the *esophagus* (ih•SAHF•uh•guhs), a long tube that leads to the stomach. Gastric juice, produced by the stomach, contains acid and chemicals that break down proteins.

After several hours in the stomach, partly digested food moves into the small intestine. Digestion of food into nutrients is completed by chemicals produced in the small intestine. Nutrients diffuse through the **villi**, projections sticking out of the walls of the small intestine, into the blood. From the small intestine, undigested food passes into the large intestine. There, water and minerals pass into the blood, and wastes are removed from the body.

Two other organs have a role in digestion. The *liver* produces bile, which is stored in the *gallbladder* until it's needed. Bile breaks down fats into smaller particles that can be more easily digested. The *pancreas* produces a fluid that neutralizes stomach acid and chemicals that help finish digestion.

✓ **In what organ of the digestive system do nutrients enter the blood?**

As food passes through the digestive system, chemicals break it down into nutrients the body cells need. ▼

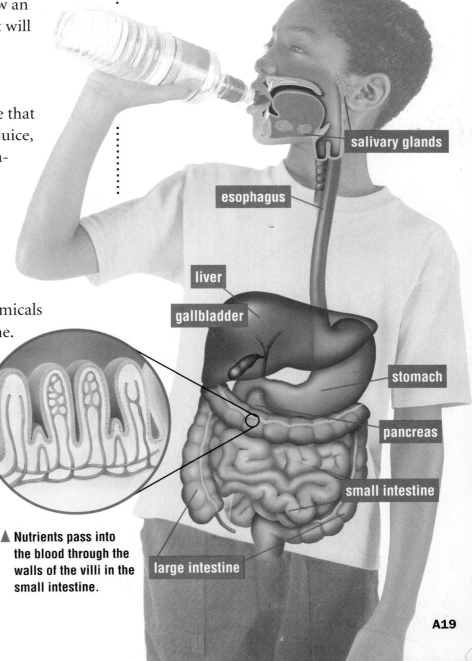

▲ Nutrients pass into the blood through the walls of the villi in the small intestine.

The Excretory System

The circulatory system supplies food and oxygen to the body's cells. It carries away waste products from the production of energy. The wastes must then be removed from the blood. This is the function of the excretory system.

Cell wastes include carbon dioxide and ammonia. As you learned, the respiratory system gets rid of carbon dioxide. Ammonia is carried by the blood to the liver, where it is changed to urea.

From the liver, urea is carried by the blood to the kidneys. These organs are located behind the liver and stomach. As blood flows through capillaries in the kidneys, urea and water enter tubes called **nephrons**.

Urine, which is urea, water, and other waste products, flows from the kidneys through tubes called *ureters*. The ureters empty into a muscular organ called the bladder. When the bladder is full, urine leaves the body through a channel called the *urethra*.

The kidneys, ureters, bladder, and urethra make up the excretory system. This system takes wastes from the blood and removes them from the body as urine. ▼

▲ Wastes and water are removed from the capillaries that run through the kidneys. Materials that the body needs are returned to the capillaries.

The excretory system keeps the amount of water in the body fairly constant.

Daily Water Gain and Water Loss in Adults

Water Gain		Water Loss	
From cell activities	400 mL	In solid wastes	100 mL
From eating	900 mL	From skin and lungs	1000 mL
From drinking	1300 mL	As urine	1500 mL
Total	2600 mL	Total	2600 mL

Cell wastes aren't the only wastes the body needs to get rid of. When you exercise, your body gets warm. Excess heat is eliminated by sweating. Sweat is a salty liquid that evaporates from the skin. Evaporation pulls heat from capillaries just below the skin. The blood and the entire body are cooled.

✓ **How do cell wastes get to the kidneys?**

Summary

Body cells are organized into tissues, organs, and systems that work together to keep the body alive. The circulatory system transports materials throughout the body. In the respiratory system, oxygen diffuses into the blood and carbon dioxide diffuses out of the blood. The digestive system breaks down food into nutrients that can be used by cells. The excretory system removes cell wastes from the blood.

Review

1. How do the functions of red blood cells and white blood cells differ?
2. In what way do the respiratory and circulatory systems work together?
3. What does the pancreas do?
4. **Critical Thinking** Between the trachea and the esophagus is a flap of skin called the *epiglottis*. What do you think its function is?
5. **Test Prep** Inhaled oxygen diffuses through the walls of the —
   A ureters
   B bronchioles
   C pulmonary arteries
   D alveoli

MATH LINK

Use Mental Math Count the number of times your heart beats in 15 sec. Then multiply the number by 4 to find your approximate heartbeat rate per minute. At that rate, how many times would your heart beat in one year?

WRITING LINK

Informative Writing—Report Find out more about one of the body systems. Write a brief report for your teacher. Describe in detail the organs of the system. Also explain how the system functions.

SOCIAL STUDIES LINK

Early Physicians Some of the first recorded information about physicians described Imhotep, who lived in Egypt during the Third Dynasty (about 2700–2500 B.C.). Find out about Imhotep and his contributions to ancient medicine.

TECHNOLOGY LINK

Learn more about the functioning of the heart by investigating *One Heart— Two Amazing Pumps* on the **Harcourt Science Explorations CD-ROM.**

LESSON 3

How Do Bones, Muscles, and Nerves Work Together?

In this lesson, you can . . .

INVESTIGATE how muscles cause movement.

LEARN ABOUT the skeletal, muscular, and nervous systems.

LINK to math, writing, art, and technology.

How Muscles Cause Movement

Activity Purpose Your respiratory, circulatory, digestive, and excretory systems function automatically, without the need for you to give them directions. Other systems in your body are under your control, at least some of the time. When you run for a bus, for example, you direct your skeleton, muscles, and nerves to work together. In this investigation you'll **observe** what your muscles do when you bend your arm.

Materials
- tape measure

Activity Procedure

1. Place your left hand on top of your right arm, between the shoulder and elbow. Bend and straighten your right arm at the elbow. **Observe** the movement by feeling the muscles in your right arm. (Picture A)

2. The muscle on the front of the upper arm is called the *biceps*. The muscle on the back of the upper arm is called the *triceps*. **Compare** the biceps and the triceps as you bend and straighten your arm. **Infer** which muscle controls the bending movement and which controls the straightening movement.

◀ In-line skating requires the coordination of your skeletal system, muscular system, and nervous system.

A22

Picture A

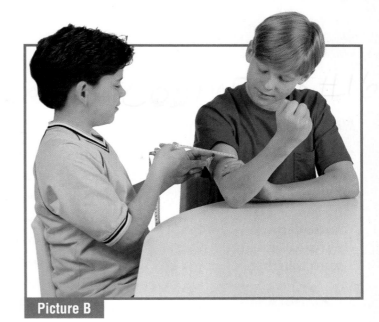
Picture B

3 Have a partner use the tape measure to **measure** the distance around your upper arm when it is straight and when it is bent. **Record** the measurements. (Picture B)

4 Repeat Steps 2 and 3, using your right hand and your left arm.

5 **Compare** the sets of measurements.

Draw Conclusions

1. What did you **infer** about the muscles controlling the bending and the straightening of your upper arm?

2. Why are two muscles needed to bend and straighten your arm? Why can't one muscle do it?

3. **Scientists at Work** Scientists often hypothesize about things they **observe**. **Hypothesize** about any differences between the measurements of your right arm and the measurements of your left arm.

Investigate Further **Plan and conduct an experiment** with different pairs of muscles. For example, try bending your leg at the knee while **observing** the muscles in your thigh. See if these measurements also support your hypothesis. **Draw conclusions** about differences in muscle sizes from the data you collected. Decide whether more data is needed to support your conclusions.

> **Process Skill Tip**
>
> When you **hypothesize**, you tell what you think will happen based on what you **observe**. A hypothesis can be tested by doing an experiment.

Systems Working Together

FIND OUT

- what the structure and function of bones are
- how the skeletal, muscular, and nervous systems work together

VOCABULARY

bone marrow
joints
tendons
ligaments
neuron
receptors

Bones and Joints

The bones of your body are living organs made up of connective tissues. The tissues include an outer protective membrane, a layer of hard material, and a soft center containing bone marrow. **Bone marrow** is a connective tissue that produces red blood cells and white blood cells.

Bone cells form canals throughout the hard layer. The cells are connected to each other and to blood vessels. Bone cells secrete the rocklike material, made of calcium, that gives bones their strength and hardness. The protective membrane surrounding bones repairs them if they are broken.

Your body contains several kinds of bones. These include long bones in your arms and legs, flat bones in your shoulders and hips, and short bones in your fingers and toes. Other kinds of bones include the irregular bones in your wrists and ankles.

Bones meet at **joints**, where they are attached to each other and to muscles. Different kinds of joints allow different kinds of movement. Hinge joints, for example, allow back-and-forth movement, like the movement of a door hinge. Ball-and-socket joints allow circular motion, like the motion of a joystick. Some joints, such as the ones in your skull, don't allow any movement.

✓ **What kind of tissue are bones made of?**

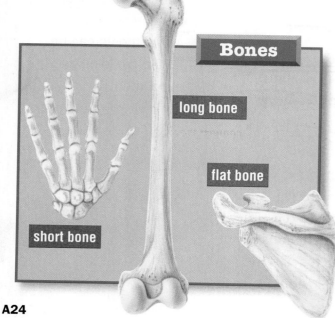

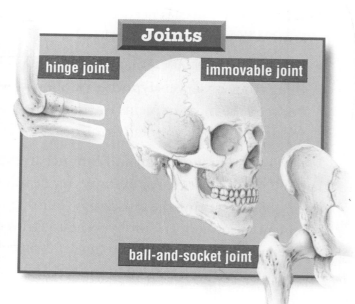

The Skeletal System

Bones are organized into a skeleton, which provides support for your body. Your skeleton also protects many of your internal organs. For example, your skull completely encloses your brain.

The outer layer of a bone provides a surface for the attachment of muscles. On each side of a joint, muscles are attached to bones by **tendons**, tough bands of connective tissue. Bones are attached to each other by **ligaments**, bands of connective tissue that hold the skeleton together.

An adult human skeleton has 206 bones. Each hand has 27 bones and each foot has 26 bones. The skull is made up of 23 bones. Not all of the skeleton is bone. Your outer ears and the tip of your nose are *cartilage*—another type of connective tissue. Cartilage also coats the ends of bones where they meet at a joint. This allows smooth movement between bones.

✓ **How are muscles attached to bones?**

The Muscular System

There are three kinds of muscles—voluntary muscles, smooth muscles, and cardiac muscles. *Voluntary muscles* move bones and hold your skeleton upright. These muscles are made up of groups of muscle tissues bound together by connective tissue. They are generally attached to two or more bones, either directly or through tendons. Where muscles attach to bones at a joint, they work in opposing pairs. As you observed in the investigation, one muscle contracts to bend a joint. Another contracts to straighten it.

Smooth muscles contract slowly and move substances through the organs they surround. These muscles run in bands around the walls of blood vessels and digestive organs.

Cardiac muscles make up the walls of the heart. Their function is to pump blood. Some cardiac muscles work together to set the heartbeat rate. They ensure that all the cardiac muscles beat at the same time.

✓ **Why must voluntary muscles work in pairs?**

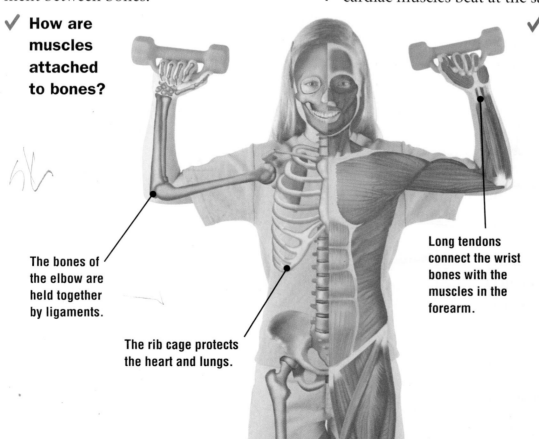

The bones of the elbow are held together by ligaments.

The rib cage protects the heart and lungs.

Long tendons connect the wrist bones with the muscles in the forearm.

The Nervous System

Your nervous system allows you to experience things and to react to your environment. It connects all the tissues and organs of your body to your brain. The nervous system consists of two parts—the central nervous system and the peripheral nervous system. The central nervous system is made up of the brain and the spinal cord. The spinal cord is a bundle of nerves, about as thick as a pencil. It runs from the base of the brain to the hips.

The peripheral nervous system consists of sensory organs, such as the eyes and ears, and body nerves. Nerves are bundles of nerve cells, or neurons. A **neuron** is a specialized cell that can receive signals and transmit them to other neurons.

Signals traveling along nerves cross the gap between neurons. This gap is called a *synapse*. The signal is carried across the synapse by chemicals produced in the sending neuron.

Sensory organs contain neurons called receptors. **Receptors** are nerve cells that detect conditions in the body's environment. Receptors in the ears detect sound waves. Those in the skin detect heat and cold, pressure, touch, and pain. Receptors in the eyes detect light and color. Those in the mouth and nose detect tastes and smells. Each receptor sends a signal through nerves to the central nervous system.

The central nervous system interprets signals it receives from nerves and determines what response is needed. Signals sent by the brain travel through nerves and direct all of the body's muscles. The brain also controls the body's automatic functions, such as respiration, circulation, and digestion.

Some muscle actions are automatic responses to situations. These are called *reflexes*. For example, when a pain signal from a skin receptor reaches the spinal cord, the nerve carrying the signal transmits it directly to a nerve that controls muscles, as well as to a nerve traveling to the brain. The reflex action of the muscles to avoid the source of pain happens before the signal reaches the brain. In other words, you react to pain before you even feel pain.

✓ **What takes place at a synapse?**

Nerves are clusters of neurons that stretch between the central nervous system—the brain and the spinal cord—and every other part of the body. ▶

spinal cord

A synapse is the gap between neurons. ▼

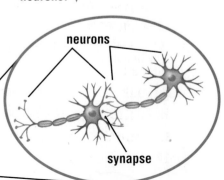

neurons

synapse

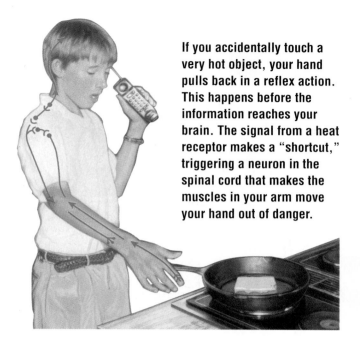

If you accidentally touch a very hot object, your hand pulls back in a reflex action. This happens before the information reaches your brain. The signal from a heat receptor makes a "shortcut," triggering a neuron in the spinal cord that makes the muscles in your arm move your hand out of danger.

Summary

Bones are living organs that make up the skeletal system. Skeletal bones move because of the action of pairs of voluntary muscles. Smooth muscles line the digestive organs and blood vessels. Cardiac muscles make up the walls of the heart. Muscles are controlled by the central nervous system. Nerves are bundles of neurons. They carry signals from sensory organs to the brain and from the brain to the muscles.

Review

1. What is produced in bone marrow?
2. What function do ligaments perform?
3. What kinds of muscles move food through the digestive system?
4. **Critical Thinking** What kind of action is a sneeze caused by pepper in the air? Explain.
5. **Test Prep** Tendons are connective tissue that —
   A makes blood cells
   B carries signals from receptors
   C connects one bone to another
   D connects a bone to a muscle

LINKS

MATH LINK

Collect/Organize Data Work with a partner to test your reaction time. Have your partner hold a meterstick vertically so that the lower end is just above the open fingers of your hand. When your partner lets the meterstick go—without warning—catch it between your fingers. Record the measurement where your fingers grasp the stick. Repeat the activity several times, and graph your results.

WRITING LINK

Informative Writing—Description Fill one bowl with ice-cold water, one with hot—but not too hot—water, and one with lukewarm water. Leave one hand in the cold water and one in the hot water for about a minute. Then put both hands into the lukewarm water. Write a paragraph describing the results.

ART LINK

Pointillism Pointillist paintings are composed of small, separate dots of color, like a photograph. The brain interprets the dots as a picture. Try making a pointillist painting of your own.

TECHNOLOGY LINK

Learn more about the benefits of exercise by viewing *The Importance of Exercise* on the **Harcourt Science Newsroom Video** in your class video library.

SCIENCE AND TECHNOLOGY

POTATO VACCINES

Scientists are using biotechnology to turn common potatoes into vaccines against deadly diseases such as cholera.

Why Potato Vaccines?

Potatoes are an inexpensive, nutritious food that most children like to eat mashed, baked, or French fried. Biologists are using genetic engineering to insert a new gene into potatoes. This gene causes the plants to produce a chemical, called a B-protein, that is a harmless part of cholera toxin, or poison. When children eat a certain amount of these genetically altered potatoes, they become vaccinated against cholera.

Potatoes are a good choice for edible vaccines because most people like them.

This white blood cell is attacking *E. coli* bacteria.

What Is Cholera?

Cholera is an infectious disease that affects about 5 million people a year, particularly in poor areas of the world. It's highly contagious, so outbreaks of cholera can easily become epidemics. Cholera toxin causes pores in cells lining the intestines to remain open when they should be closed. Water diffuses from the blood into the intestines and then out of the body as diarrhea. People with cholera can lose so much water so quickly that they become very sick and may even die.

How Does the Potato Vaccine Work?

Scientists at the Loma Linda University School of Medicine in California have added the gene that produces the cholera toxin's B-protein to the genes of potato plants. The B-protein attaches to cells in the intestines and triggers the production of antibodies against cholera.

Scientists tested the potato vaccines by feeding genetically altered potatoes to mice. Then they examined tissue from the mice's intestines to see what happened when it was exposed to cholera toxin. They found that only half as much water passed through this tissue compared with tissue from mice that had not eaten the altered potatoes.

People seldom eat potatoes raw, so scientists had to make sure that cooking wouldn't destroy the vaccine. They found that after cooking, the genetically altered potatoes had about half the vaccine they started with. But this is still enough vaccine to be effective. One cooked potato a week for one month provides enough cholera toxin B-protein to protect against the disease for years.

Other uses of the potato vaccine

Toxins produced by the cholera bacteria are nearly identical to toxins produced by another dangerous intestinal bacteria called *E. coli*. So the potato vaccine that works against cholera may also work against *E. coli* toxins. Scientists are also looking for ways to improve the potato vaccine so that it will help destroy the bacteria directly, not just the toxins the bacteria produce.

THINK ABOUT IT

1. Why do you think potato vaccines would be easier to distribute and give to people than injected vaccines?
2. What would be an advantage of inserting vaccine-producing genes into foods that are eaten raw, such as bananas?

CAREERS
GENETICIST

What They Do
Geneticists study the ways genes and chromosomes combine to produce variety in plants and animals.

Education and Training
People wishing to become geneticists should earn an M.D. degree, or a Ph.D. degree in cell biology.

WEB LINK
For Science and Technology updates, visit the Harcourt Internet site.
www.harcourtschool.com

PEOPLE IN SCIENCE

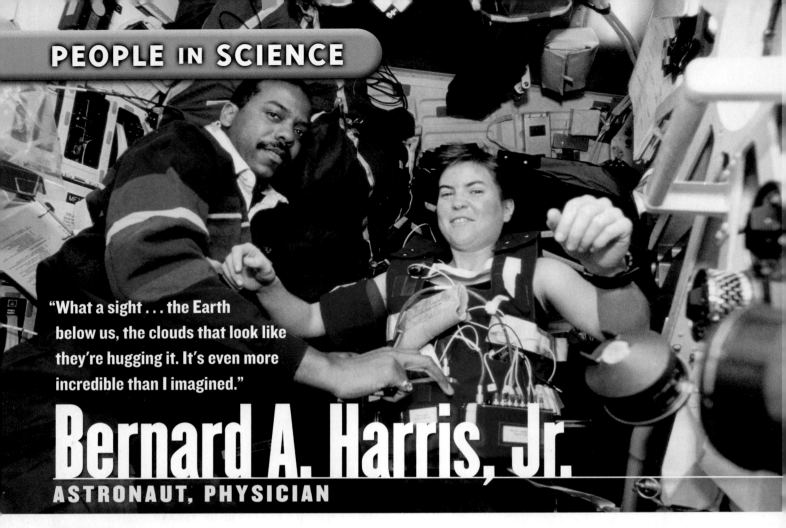

"What a sight... the Earth below us, the clouds that look like they're hugging it. It's even more incredible than I imagined."

Bernard A. Harris, Jr.
ASTRONAUT, PHYSICIAN

When Bernard A. Harris, Jr., opened the hatch of the space shuttle *Discovery* and stepped outside, he became the first African American ever to walk in space. It was an amazing moment for a man who, from the age of 13, had dreamed of becoming an astronaut. His dream had been born as, watching on television, he saw Neil Armstrong take that first step on the moon. Although he kept his goal a secret, Dr. Harris worked constantly toward it, studying biology and medicine.

Dr. Harris's first trip in space was aboard the space shuttle *Columbia* in 1993. He and his crewmates experimented to find out the effects of space travel on the human body. The researchers learned that for every month astronauts spend in space, they lose 22.4–44.5 newtons (5–10 lb) of their body weight. In addition, they lose 20 percent of their blood volume because their bone marrow makes less blood.

While in space, Dr. Harris used sound waves to "see" his own heart and discovered that, under the conditions of near weightlessness, it shrank and shifted in his chest. "I had to listen for it in a slightly different place," he recalls.

Today, Dr. Harris is vice-president of SPACE-HAB, Inc., a company that furnishes payloads and experimental modules for NASA's space shuttles. He is also active in the STARS program (Space Technology and Research Students). This is a program that enables students to participate in experiments that are conducted in space.

THINK ABOUT IT

1. Why did Dr. Harris have to listen for his heart in a different place?
2. What physical effects might astronauts experience if they remain in space for months?

ACTIVITIES FOR HOME OR SCHOOL

BALLOON LUNGS

How do lungs work?

Materials
- 2 balloons
- scissors
- plastic soda bottle

Procedure

1. Remove the cap and cut the bottom off the bottle.

2. Put one balloon into the bottle. Secure the lip of the balloon to the top of the bottle.

3. Cut the lip off the second balloon. Stretch the large part of the second balloon over the bottom of the bottle.

4. With your fingers, pull down on the second balloon and then release it. Observe what happens to the first balloon.

Draw Conclusions

When you pull on the second balloon, what happens inside the bottle? What part of the respiratory system does each part of your model represent?

SKELETAL SYSTEMS

What adaptations do skeletons show?

Materials
- butcher paper
- 5 people
- meterstick
- marker

Procedure

1. Measure out about 7 m of butcher paper.

2. Have one person lie in the center of the paper. This person should stretch out his or her arms as shown.

3. Have two other people lie end-to-end on each side of the first person.

4. Use the marker to draw around the first person, including the top edge of his or her outstretched arms and thumbs.

5. To complete the top edge, draw a sloping line the lengths of the people on both sides as shown. Draw the lower edge with four points and four scallops as shown.

Draw Conclusions

While the skeletal systems of all mammals are similar, there are differences due to various adaptations. The skeletons of bats show adaptations for flight. If humans could fly, how many times longer than their bodies would their wings need to be? What other skeletal adaptations of mammals can you think of?

CHAPTER 1 Review and Test Preparation

Vocabulary Review

Use the terms below to complete the sentences. The page numbers in ( ) tell you where to look in the chapter if you need help.

cell (A6)
cell membrane (A8)
nucleus (A8)
cytoplasm (A9)
diffusion (A10)
osmosis (A10)
tissue (A12)
organ (A12)
system (A12)
capillaries (A17)
alveoli (A18)
villi (A19)
nephron (A20)
bone marrow (A24)
joints (A24)
tendons (A25)
ligaments (A25)
neuron (A26)
receptors (A26)

1. The ___ are projections of the inside wall of the small intestine.
2. The smallest blood vessels are the ___.
3. Similar cells work together in a ___, which is part of an ___, which is part of a ___.
4. Cells that detect conditions in the body's environment are ___.
5. The ___ determines cell activities, and the ___ regulates what enters and leaves the cell.
6. Air sacs in the lungs through which oxygen and carbon dioxide diffuse into and out of the blood are the ___.
7. The ___ is the basic unit of structure of all living things.
8. At ___, bones are connected to each other by ___, while ___ connect bones to muscles.
9. A ___ is a cell that can receive and transmit signals.
10. The structure in the kidney that filters urea and water from the blood is a ___.
11. Particles move from areas where there are a lot of them to areas where there are fewer of them by ___. Water and materials dissolved in water pass through a membrane by ___.
12. The jellylike substance between the cell membrane and the nucleus is the ___.
13. Blood cells are produced in ___.

Connect Concepts

Complete the chart by filling in terms from the Word Bank. Some terms are used more than once.

smooth muscle
oxygen
small intestine
nutrients
cardiac muscle
lungs
esophagus
alveoli
voluntary muscle

Food
travels down the **14.** to the stomach and then to the **15.**, where **16.** pass into the blood.
Air
travels down the trachea to the **17.**, where **18.** diffuses into the blood through the walls of the **19.**
16. and **18.**
are carried by the blood to muscle cells. The energy they produce in **20.** tissue is used to move bones. The energy they produce in **21.** tissue is used to move food through the digestive system. The energy they produce in **22.** tissue is used to move blood through the circulatory system.

A32

Check Understanding
Write the letter of the best choice.

23. The tissue that makes up bones, tendons, and ligaments is —
    A connective tissue
    B epithelial tissue
    C muscle tissue
    D voluntary tissue

24. Bile, produced by the liver, breaks down —
    F muscle tissue
    G fats
    H stomach acid
    J white blood cells

25. The urea and water that make up urine are removed from the blood in the —
    A bladder
    B kidneys
    C pancreas
    D urethra

26. A plant stands up straight because of water pressure against the —
    F cell walls
    G nucleus
    H cellular membrane
    J chloroplasts

27. Neurons transmit signals across a —
    A synapse
    B nucleus
    C cell wall
    D receptor

28. The liver is an organ of the —
    F digestive system only
    G excretory and digestive systems
    H circulatory system only
    J circulatory and excretory systems

Critical Thinking

29. A nerve can carry signals both to the brain and from the brain. Why can't an individual neuron do this?

30. Platelets in the blood cause clotting. Why is this important?

31. On each side of the heart, a valve allows blood to travel from the upper chamber to the lower chamber. Why is this important?

Process Skills Review

32. If you are **observing** cells under a microscope, what will lead you to **infer** that the cells are animal cells?

33. How would you test the **hypothesis** that there is never a ball-and-socket joint between a hinged joint and the end of a limb of a skeleton?

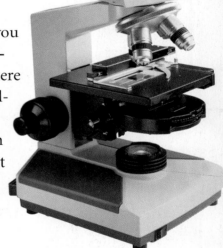

Performance Assessment
Body Systems

Use colored pencils to draw the organs and vessels of the excretory system on a human body outline. Label each part. Describe how this system works to eliminate waste from the body cells.

A33

CHAPTER 2
Classifying Living Things

Vocabulary Preview

classification
kingdom
moneran
protist
fungi
genus
species
vertebrates
mammals
reptiles
amphibians
fish
birds
invertebrates
vascular plants
nonvascular plants

Have you ever noticed how some living things have the same kinds of parts? A cat and a dog each have fur, four legs, and a tail. An ant and a cockroach each have six legs. Scientists look at the similarities among living things and put them into groups.

Fast Fact

A kelp may look like a plant, but it's not. The leaflike and stemlike parts of a kelp are different from those of plants. Instead, kelps belong to the protist kingdom. Other protists include microscopic amoebas and paramecia.

Fast Fact

A duckbill platypus looks like a mixed-up animal. You might think it's a bird because it has a bill and lays eggs. But scientists say it's a mammal, like a dog or cat, because it has fur and produces milk for its young.

Fast Fact

This Ithaca bog beetle was in an insect collection at Cornell University in Ithaca, New York, for 85 years before anyone realized it had no scientific name. Scientists think there are many more living things to be classified and named. Many haven't even been discovered yet.

Numbers of Living Things

Type of Living Thing	Number of Known Species, or Kinds
Insects	1,000,000
Fish	30,000
Orchids	20,000

LESSON 1

How Do Scientists Classify Living Things?

In this lesson, you can . . .

INVESTIGATE ways to group nonliving objects.

LEARN ABOUT classification.

LINK to math, writing, art, and technology.

Classifying Shoes

Activity Purpose You've probably looked for a certain book in a library. Imagine how hard it would be to find a book in a library full of books if they were not grouped by topic. In this activity you will practice **classifying** some familiar items.

Materials
- shoes
- newspaper or paper towels

Activity Procedure

1. Take off one shoe and put it with your classmates' shoes. If you put the shoes on a desk or table, cover it first with newspaper or paper towels. (Picture A)

2. Find a way to **classify** the shoes. Begin by finding two or three large groups of shoes that are alike. Write a description of each group. (Picture B)

3. **Classify** the large groups of shoes into smaller and smaller groups. Each smaller group should be alike in some way.

◀ This sun bear is a type of animal called a mammal. What other types of animals can you name?

A36

Picture A

Picture B

4. Write a description of each smaller group.
5. Stop classifying when you have sorted all the shoes into groups with two or fewer members.

Draw Conclusions

1. What features did you use to **classify** the shoes?
2. **Compare** your classification system with a classmate's system. How are your systems alike? How are they different?
3. **Scientists at Work** Scientists **classify** living things to show how living things are alike. Why might it be important for scientists to agree on a set of rules for classifying living things?

Investigate Further **Classify** other groups of things such as toys, cars, or pictures of plants and animals. Write a brief explanation of your classification system.

Process Skill Tip

When you **classify** things, you put them into groups based on how they are alike. Things that are not similar are in different groups. Things that have similar characteristics form a group.

LEARN ABOUT

Classification

Grouping Living Things

If you were asked to go to the grocery store to buy fresh peaches, how would you find them? You know how your grocery store is set up, so you would probably go to the produce department and find the fruit section. There you would look for peaches. If a store put some fruit with the cereal and some with the meat, finding peaches would be much more difficult.

Like grocery shoppers, scientists need to be able to find things easily. Just as you did with shoes in the investigation, scientists look at living things and identify their characteristics. They then group together living things that have similar features. This act of grouping things by using a set of rules is called **classification** (klas•uh•fih•KAY•shuhn).

✓ **How do scientists group living things?**

FIND OUT

- why scientists group living things
- the names of the five largest groups of living things

VOCABULARY

classification
kingdom
moneran
protist
fungi
genus
species

The living things shown in the forest scene and in the smaller photos belong to several different groups.

Bacterium

Water strider

Paramecium

The Five Kingdoms

Kingdom	Important Characteristics	Examples
Animals	Many-celled, feed on living or once-living things	Monkeys, birds, frogs, fish, spiders
Plants	Many-celled, make their own food	Trees, flowers, grasses, ferns, mosses
Fungi	Most many-celled, absorb food from other living things or dead things such as logs	Mushrooms, yeasts, molds
Protists	Most one-celled, make their own food or feed on living or once-living things	Algae, amoebas, diatoms
Monerans	One-celled, no cell nuclei, some make their own food, some feed on living or once-living things	Bacteria

Grouping by Similarities and Differences

Scientists classify for many reasons. Classifying living things makes it easier to find and share information about them. When scientists discover a new living thing, classification can show how the new living thing relates to others that are already classified.

All living things can be classified into one of five kingdoms. A **kingdom** (KING•duhm) is the largest group into which living things can be classified. Every member of a kingdom has some characteristics that are the same as those of other members. For example, bacteria are monerans. Every member of the **moneran** (muh•NER•uhn) kingdom has only one cell. The cell has no nucleus.

Compare the bacterium to the paramecium on page A38. Paramecia are protists. Most members of the **protist** (PROH•tist) kingdom also have only one cell. However, each cell does have a nucleus.

Fungi make up a third kingdom. **Fungi** (FUHN•jy) have nuclei, and most are many-celled. In some ways they are like plants, but they can't make their own food as plants do. You have eaten fungi if you've ever eaten mushrooms.

Plants and animals make up the other two kingdoms. Every day you see members of these two kingdoms, such as grass, flowers, cats, and dogs.

✓ **How are monerans and protists the same? How are they different?**

Forming Smaller Groups

Classification doesn't stop at the kingdom level. Scientists study the living things in each kingdom to see how they are alike and how they are different. They use characteristics to make smaller and smaller groups, and they give each smaller group a name. The most specific classification groups have only one type of living thing. The chart below shows how brown bears can be classified by using this method.

Most living things have a common name such as *brown bear*. But common names may be different in different places. It's important to have names that scientists everywhere recognize. For this reason, scientists name animals with the labels of the two smallest classification groups. The name of the second smallest group, the **genus** (JEE•nuhs), is joined with the name of the smallest group, the **species** (SPEE•sheez). For example, the scientific name for a house cat is *Felis domesticus*, and a brown bear is called *Ursus arctos*.

✓ **How do scientists form smaller groups of living things?**

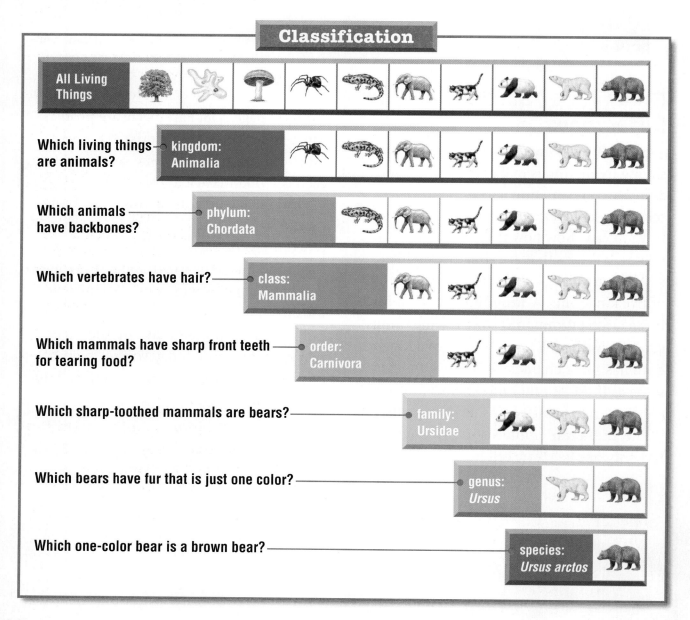

▲ This bear cub's scientific name is *Ursus* (genus) *arctos* (species).

Summary

Scientists organize living things to make studying and discussing them easier. Scientists classify all living things into five kingdoms—monerans, protists, fungi, plants, and animals. The five kingdoms are divided into smaller groups.

Review

1. Why do scientists organize information about living things?
2. What are the five kingdoms of living things?
3. How do scientists name each type of living thing?
4. **Critical Thinking** There are probably millions of living things that scientists haven't discovered yet. If scientists were to find a living thing that didn't fit into any of the five kingdoms, what would they need to do?
5. **Test Prep** Which kingdom contains one-celled living things without nuclei?
   A plants
   B monerans
   C fungi
   D protists

LINKS

MATH LINK

Display Data Suppose you have found a cave in which animals—three snakes, six bats, and a bear—are living. To report your discovery to your classmates, make a bar graph showing the types and numbers of animals in the cave.

WRITING LINK

Informative Writing—Description Suppose you've discovered a new species of living thing. For your teacher, write two or three paragraphs to describe how you found it, what its characteristics are, and how you decided on its name.

ART LINK

Designing Labels Think about how you could improve the organization of your books, games, or CDs. Classify them and then design picture labels for each group. Put the labels on your books, games, or CDs so that you can more easily find the ones you want.

TECHNOLOGY LINK

Learn more about some bears by viewing *China Panda* on the **Harcourt Science Newsroom Video** in your classroom video library.

LESSON 2

How Are Animals Classified?

In this lesson, you can . . .

 INVESTIGATE a model of a backbone.

 LEARN ABOUT vertebrate classification.

 LINK to math, writing, health, and technology.

Building a Model Backbone

Activity Purpose Animals are classified into several groups. Animals in one of these groups all have a backbone that protects the spinal cord and helps support the body. In this investigation you will **make and use a model** backbone.

Materials
- chenille stem
- wagon-wheel pasta, uncooked
- candy gelatin rings

Activity Procedure

1 CAUTION **Never eat anything you use in an Investigate.** Bend one end of the chenille stem. Thread six pieces of wagon-wheel pasta onto the stem. Push the pasta down to the bend in the stem. Bend the stem above the pasta to hold the pasta in place.

2 Bend and twist the stem. What do you see and hear?

3 Take all the pasta off the chenille stem except one. Thread a candy gelatin ring onto the stem, and push it down. (Picture A)

◀ Birds have a backbone. Worms, like the one the bird is eating, don't.

Picture A

Picture B

4 Add pasta and rings until the stem is almost full. Bend the stem above the pasta and rings to hold them in place. (Picture B)

5 Bend and twist the stem. What do you see and hear?

6 Draw pictures of the model backbones you made. **Compare** your models with that shown in the picture on page A57.

Draw Conclusions

1. A real backbone is made of bones called vertebrae (VER•tuh•bray) and soft discs that surround the spinal cord. What does each part of your final model stand for?

2. How is your final model like a real backbone?

3. Study your final model again. What do the soft discs do?

4. **Scientists at Work** Scientists use models to study how things work. **Make a model** to test this **hypothesis:** A piece of dry, uncooked spaghetti or some other material would work better than a chenille stem to stand for the spinal cord in a model. **Experiment** to see. Then write a report of your experiment. Be sure to include the results of any tests you conducted with other materials, and any conclusions you drew about using those materials in a model backbone.

Process Skill Tip

Some objects are too big, too small, or too far away to observe directly. You can't observe your backbone directly because it is inside your body. But you can **make a model** to learn more about it.

LEARN ABOUT

FIND OUT
- how vertebrates are classified
- what groups of animals make up vertebrates

VOCABULARY
vertebrate
mammal
reptile
amphibian
fish
bird
invertebrate

Animal Classification

Animals with a Backbone

You are probably familiar with many members of the animal kingdom. An animal is a living thing made up of many cells that have nuclei. Animals can't make their own food. They must eat to survive. Scientists divide the animal kingdom into two large groups. One group of animals has backbones.

Animals that have a backbone are called **vertebrates** (VER•tuh•brits). Most vertebrates have sharp senses and large brains. These characteristics help them survive in their surroundings.

The large group of vertebrates is divided into several smaller groups. **Mammals** (MAM•uhlz) have hair and produce milk for their young. Cats and dogs are mammals that you may have as pets. Lizards, snakes, and turtles are reptiles. **Reptiles** (REP•tylz) have dry, scaly skin. **Amphibians** (am•FIB•ee•uhnz) have moist skin and no scales. Most of them begin life in water, but they live on land as adults. Frogs, toads, and newts are amphibians.

Sharks, eels, bass, and tuna are fish. **Fish** are vertebrates that live their entire lives in water. Most fish have hard scales covering their bodies and gills to take the oxygen they need directly from the water.

Mongoose

Frog

◄ A mongoose and a frog are both vertebrates. What characteristic do they share with a snake?

Snake skeleton

▲ Although a shark's backbone isn't made of bone, a shark is still a vertebrate.

Birds are vertebrates with feathers. A bird's feathers keep it warm and help it to fly. Owls, robins, and parrots are birds. Some birds, such as penguins, don't fly.

✓ **What characteristic do all vertebrates have in common?**

Only a small part of all the animals in the world have a backbone. ▼

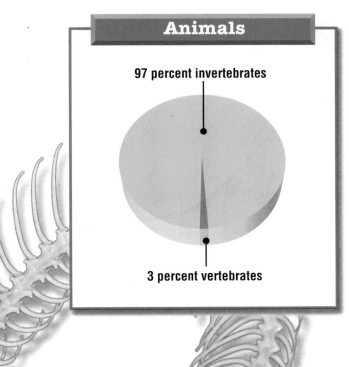

Animals

97 percent invertebrates

3 percent vertebrates

Snail

Tortoiseshell beetle

Sea sponge

Snails, sea sponges, and beetles are invertebrates. None of these animals has a backbone.

Animals Without a Backbone

Animals without a backbone are called **invertebrates** (in•VER•tuh•brits). There are many more types of invertebrates than types of vertebrates. Most invertebrates are smaller than vertebrates.

Arthropods (AR•throh•pahdz) are invertebrates with legs that have several joints. Their bodies have two or more parts, and they often have an outer covering that protects them. There are several groups of arthropods. Insects make up the largest group. Adult insects, such as beetles and bees, have six legs. Spiders aren't insects. They and other arthropods, such as mites, horseshoe crabs, and scorpions, have eight legs.

Mollusks (MAHL•uhsks) are invertebrates that may or may not have a hard outer shell. Snails, clams, and squids are mollusks.

Invertebrates also include several groups of *worms*. Worms have no shells, legs, or eyes. Earthworms, tapeworms, and flatworms belong to different groups of invertebrates.

✓ **What characteristic do all invertebrates have in common?**

A45

THE INSIDE STORY

Body Parts for Jumping

Frogs and grasshoppers are in different animal groups. Both animals are known for their ability to jump. Their back legs are different, but they work in much the same way.

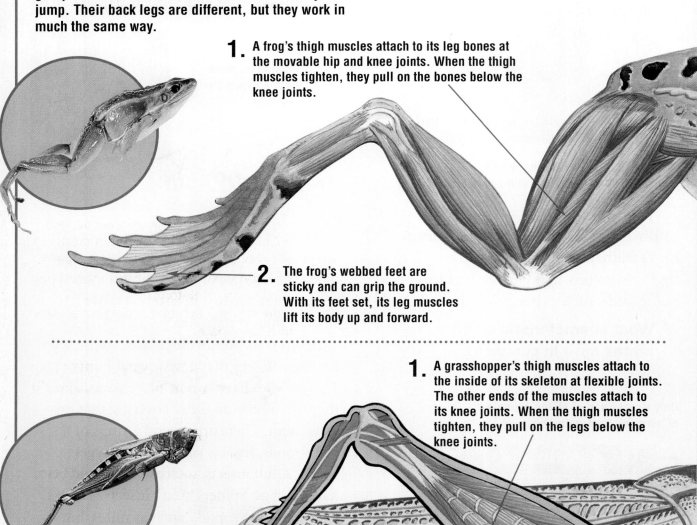

1. A frog's thigh muscles attach to its leg bones at the movable hip and knee joints. When the thigh muscles tighten, they pull on the bones below the knee joints.

2. The frog's webbed feet are sticky and can grip the ground. With its feet set, its leg muscles lift its body up and forward.

1. A grasshopper's thigh muscles attach to the inside of its skeleton at flexible joints. The other ends of the muscles attach to its knee joints. When the thigh muscles tighten, they pull on the legs below the knee joints.

2. The grasshopper's feet have claws and grip the ground. With its feet set, its leg muscles lift its body up and forward.

A Closer Look at Animals

Not all animals have a backbone, but most animals have skeletons and muscles that work together to allow the animals to move. The skeletons of vertebrates are made up of bones that support their bodies from the inside. Muscles attach to the bones at movable joints.

Most invertebrates have skeletons that form hard outer coverings. These skeletons are made of a material much like human fingernails. Muscles attach on the inside of these coverings at flexible joints.

✓ **Where do muscles attach to the skeletons of animals?**

Summary

Vertebrates, such as mammals, reptiles, amphibians, birds, and fish, have backbones. Invertebrate animals, such as arthropods, mollusks, and worms, do not have backbones.

Review

1. Which group of vertebrates begins life in water and later lives on land?
2. How is a spider different from an insect?
3. How are the skeletons of vertebrates and invertebrates different?
4. **Critical Thinking** How might having sharp senses and large brains help vertebrates survive?
5. **Test Prep** Which animals are **NOT** vertebrates?
   - A reptiles
   - B mammals
   - C amphibians
   - D arthropods

MATH LINK

Display Data Vertebrate skeletons are made up of bones. The adult human spine has 33 bones. Find out how many bones the spines of five other vertebrates have. Make a bar graph to show what you learn.

WRITING LINK

Informative Writing—Explanation You've learned that skeletons support animals' bodies and help them move. Skeletons also protect animals' organs. Would you prefer to have a hard outer shell or the skeleton you have now? Write a paragraph to explain your answer to a classmate.

HEALTH LINK

Prevention Calcium helps build strong bones. Eating calcium-rich foods helps prevent bone problems as you get older. Find out which foods are rich in calcium. Then make a chart to post in your kitchen at home.

TECHNOLOGY LINK

Learn more about vertebrates by investigating *Vertebrate Challenge* on the **Harcourt Science Explorations CD-ROM.**

LESSON 3

How Are Plants Classified?

In this lesson, you can . . .

INVESTIGATE plant stems.

LEARN ABOUT plant classification.

LINK to math, writing, literature, and technology.

Plant Stems

Activity Purpose You have learned that animals can be classified by whether they have a backbone. Plants also can be classified by their parts. One of those parts is the stem. In this investigation you will **observe** a stalk of celery to help you **infer** what stems do. Although a celery stalk is actually a celery leaf, it acts like a stem.

Materials
- fresh celery stalk with leaf blades
- plastic knife
- two containers
- water
- red food coloring
- blue food coloring
- paper towels
- hand lens

Activity Procedure

1. Use the plastic knife to trim the end off the celery stalk. Split the celery from the middle of the stalk to the bottom. Do not cut the stalk completely in half. (Picture A)

2. Make a chart like the one here.

Time	Observations

▼ These flowers and mosses are two different types of plants. They move water in different ways.

Picture A

Picture B

3. Half-fill each container with water. Add 15 drops of red food coloring to one container. Add 15 drops of blue food coloring to the other container.

4. With the containers side by side, place one part of the celery stalk in each container of colored water. You may need to prop the stalk up so the containers don't tip over. (Picture B)

5. **Observe** the celery every 15 minutes for an hour. **Record** your observations on your chart.

6. After you have completed your chart, put a paper towel on your desk. Take the celery out of the water. Cut about 2 cm off the bottom of the stalk. Use the hand lens to **observe** the pieces of stalk and the freshly cut end of the stalk.

Draw Conclusions

1. Where did the water travel? How do you know?
2. How fast did the water travel? How do you know?
3. **Scientists at Work** Scientists **infer** what happens in nature by making careful observations. Based on this investigation, what can you infer about the function of stems?

Investigate Further **Hypothesize** about how you could change a white carnation into a flower with two colors. **Plan and conduct an experiment** to test your hypothesis.

Process Skill Tip

When you **infer**, you use what you observe to explain what happened. Inferring is like using clues to solve a mystery. Observing carefully, like finding good clues, can help you infer correctly.

A49

Plant Classification

Plants with Tubes

FIND OUT

- how the plant kingdom is divided
- members of each main group of plants

VOCABULARY

vascular plant ✓
nonvascular plant

All plants are members of the plant kingdom. Plants have many cells, and their cells have nuclei. Unlike animals, plants do not need to eat other living things to survive. Instead, they make their own food. Scientists divide the plant kingdom into two main groups. One group of plants has tubes. The other group does not.

Vascular (VAS•kyuh•ler) **plants** have tubes. These tubes can be found in roots, stems, and leaves. Water and nutrients enter a plant through the roots. The tubes in the roots then carry this mixture to the stems. You observed some stemlike tubes in the investigation. Tubes in stems carry the water and nutrients to tubes in a plant's leaves. A different set of tubes carries the food the leaves make to the other parts of the plant. Some food tubes run from the leaves to the roots.

Ferns are a type of vascular plant. The tubes of fern stems form a network. They often split apart and rejoin. Cells that make up the tubes are stiff. This helps provide support for the fern as its stems grow.

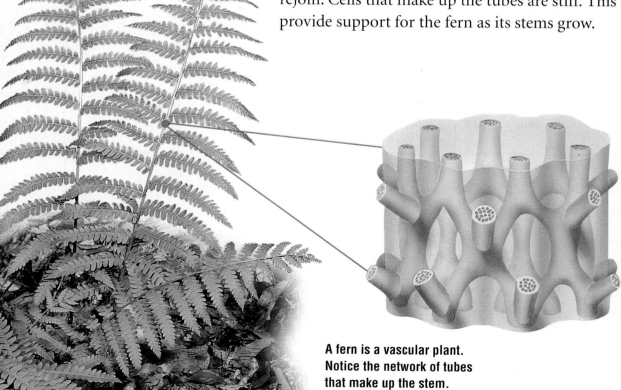

A fern is a vascular plant. Notice the network of tubes that make up the stem.

Sapwood
Growth ring
Heartwood
Bark

Trees are another type of vascular plant. The stems of trees contain cells that are woody, or very hard. Most large bushes also have woody stems. The largest woody stem of a tree is called the trunk. Look at the detailed slice of the tree trunk. The center of the trunk is made of clogged tubes called *heartwood*. Around the heartwood is a ring of *sapwood*. The sapwood tubes carry water and nutrients. Each year, a new set of tubes forms, and an old set dies, adding a growth ring to the trunk. The outside layer of the trunk is called the bark. The bark contains living tubes that carry food, and dead cells that protect the trunk.

There are many other types of vascular plants. Any plant that has flowers or cones is a vascular plant.

✓ **What is carried by the tubes of vascular plants?**

The giant sequoia is a conifer. Conifers (KAHN•uh•ferz) are vascular plants that produce cones. As new sapwood is added each year, a new growth ring forms in the trunk. ▶

Plants Without Tubes

Have you ever seen something that looked like a green carpet growing on stones and walkways? If so, you probably saw moss. Moss is a nonvascular plant. **Nonvascular** (nahn•VAS•kyuh•ler) **plants** don't have tubes. Water must soak into the plants and pass slowly from cell to cell. Food made in the plants must travel with the water from cell to cell. For this reason, nonvascular plants live in damp places and don't grow to be large or tall.

Mosses are often the first plants to grow on bare rock. Their rootlike structures help break down the rock into soil. When the mosses die, their dead bodies help to enrich the soil, making it more fertile. Nonvascular plants need fertile, moist soil in which to grow.

Nonvascular plants have no roots, stems, or leaves. The lobes, or rounded parts, of the liverwort may look like leaves, but they are not true leaves because they have no tubes.

✓ How does water travel through a nonvascular plant?

▲ Moss grows in shady, damp places.

▲ These plants are liverworts. Liverworts also grow in damp places.

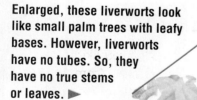

Enlarged, these liverworts look like small palm trees with leafy bases. However, liverworts have no tubes. So, they have no true stems or leaves. ▶

Summary

Scientists have classified plants into two main groups. Vascular plants, such as ferns and trees, have tubes. Because they have tubes to carry water, nutrients, and food, vascular plants can grow quite tall.

Nonvascular plants, such as mosses, do not have tubes. So water must move from cell to cell. These plants need to live in a moist place, and they do not grow to be very large.

Review

1. What are the two main groups of plants?
2. Where are the tubes of vascular plants found?
3. Because nonvascular plants do not have tubes, in what kind of place do they need to grow?
4. **Critical Thinking** What probably would happen to a plant if its main stem were crushed or broken?
5. **Test Prep** Which of these is an example of a nonvascular plant?
    A conifer   C moss
    B fern   D flower

MATH LINK

Collect/Organize/Display Data The width of a growth ring depends on the amount of rainfall the tree received that year. Wide rings form in rainy years. Narrow rings form in dry years. Examine a tree stump or the end of a log. Count the growth rings, and then measure the width of each ring. Make a line graph or a bar graph to show what you see. What can you infer from your graph?

WRITING LINK

Informative Writing—Description Gather several types of plants, and examine their characteristics. Write clues describing each plant. Your clues can be about color, smell, height, size, or the plant's use, or they may tell where it was found. Read your clues to your classmates, and see if they can guess your plant.

LITERATURE LINK

Sugaring Time Would you like to learn how maple syrup is made from sap that flows through the tubes in maple trees? Read *Sugaring Time* by Kathryn Lasky.

GO ONLINE TECHNOLOGY LINK

Learn more about plant classification by visiting the Harcourt Learning Site for related links, activities, and resources.
www.harcourtschool.com

SCIENCE THROUGH TIME

Naming Living Things

Marmota monax (gopher or woodchuck)

People have classified living things for a long time. Cave people probably sorted animals into groups such as those that were good to eat and those that were likely to eat you. Classification is an important first step in the study of almost anything.

The history of classification shows how ideas in science can change through time. As scientists learn more, they change their ideas about how things work. For example, the first recorded classification system for living things that we know about was developed by Aristotle. Aristotle was a philosopher,

Gopherus polyphemus (gopher tortoise)

In different parts of the United States, both of these animals are called gophers. The scientific name of each helps you know to which animal a scientist is referring.

teacher, and scientist in ancient Greece. In about 350 B.C. he classified living things into two large groups—plants and animals. He divided animals by how they looked, how they behaved, and where they lived. He divided plants by their size and shape. He said that the three main divisions of the plant kingdom were trees, shrubs, and herbs (small plants such as grasses).

A New System

Aristotle's system didn't work for all plants and animals. However, it was used for more than 2000 years. In 1753 Carolus Linnaeus published the system that is the basis for the system we use today. Linnaeus, like Aristotle, divided living things into two kingdoms. However, for more exact sorting, he then broke the kingdoms into many smaller groups. The smallest group is the species. Today, scientists use genus and species names to identify living things.

Which Cat?

Genus and species names are important because they help scientists and other people talk about exactly the same organisms. For example, people use different names for one kind of large cat—*puma*, *panther*, and *mountain lion*. However, a scientist uses only the name *Felis concolor*. That way you know exactly which type of cat he or she is talking about.

As microscopes and other instruments for the study of living things became better, people began to realize that there were probably more than two kingdoms. Fungi were the first organisms classified as a new kingdom. After a lot of study, living things were divided into five different kingdoms. Some scientists now suggest that there may be as many as seven kingdoms.

As we learn more, our ideas about how living things are related change. As those ideas change, the way we classify the world of living things also changes. Each change in the classification system is a direct result of more study and better understanding of the relationships of living things.

THINK ABOUT IT

1. Linnaeus classified living things. Give examples of two other classification systems and what they classify.
2. How have changes in technology affected the classification of living things?

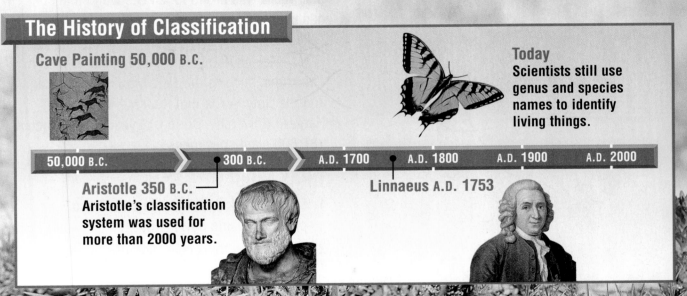

The History of Classification

Cave Painting 50,000 B.C.

Today Scientists still use genus and species names to identify living things.

50,000 B.C. — 300 B.C. — A.D. 1700 — A.D. 1800 — A.D. 1900 — A.D. 2000

Aristotle 350 B.C. Aristotle's classification system was used for more than 2000 years.

Linnaeus A.D. 1753

PEOPLE IN SCIENCE

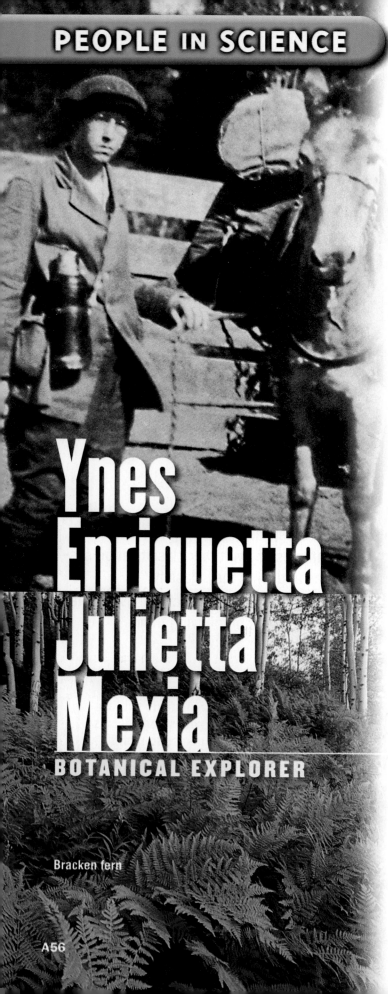

Bracken fern

Ynes Enriquetta Julietta Mexia
BOTANICAL EXPLORER

Ynes Mexia spent the last 13 years of her life, from 1925 to 1938, collecting plant specimens outside the United States. She was the daughter of an agent for the Mexican government, so she knew other languages and understood other cultures. This helped her when she traveled. She visited many places, including Mexico, Alaska, Brazil, Ecuador, Argentina, Bolivia, and Peru. During her trips she discovered almost 50 new plant species.

While living in San Francisco, Ms. Mexia traveled with the local Sierra Club. She took classes in natural science at the University of California and became interested in botany (BAHT•uhn•ee). She took a class on flowering plants, and it changed her life.

The botany class led to her first collecting trip, with botanist Roxanna S. Ferris. The trip was cut short when Ms. Mexia fell from a cliff. She broke several ribs and injured her hand. But before her fall, she had already collected 500 species of plants. One new species was named in Ms. Mexia's honor.

Nearly all Ms. Mexia's trips were to tropical countries. Because of the humid climates, it was difficult to dry and preserve plant samples. Alice Eastwood, a noted botanist, taught Ms. Mexia how to collect and preserve plants. Later, Ms. Mexia was proud to tell Eastwood that she had been able to preserve every specimen she collected!

Ms. Mexia's samples went to important museums, such as the Field Museum in Chicago and the Gray Herbarium (her•BAIR•ee•uhm) at Harvard University. During 13 years she collected 137,600 plant specimens.

THINK ABOUT IT

1. Do you think it is easier to collect and preserve plants now? Why?
2. How do you think collecting plants helps scientists understand more about plant classification?

ACTIVITIES FOR HOME OR SCHOOL

BACKBONE CONSTRUCTION

How do backbones give vertebrates flexible support?

Materials
- construction paper
- tape
- scissors
- book

Procedure

1. Roll the paper into a tube about 5 cm across. Tape all along the edge.

2. Stand the tube on one end. Will the tube hold up a pair of scissors? A book? More than one book?

3. Squeeze the tube gently to make an oval. Make slits about 2 cm apart all down the tube. Cut the slits from each side almost to the middle.

4. Experiment to see how much weight the tube will now hold up.

Draw Conclusions

What happened each time the tube gave way? How did the cuts change the tube?

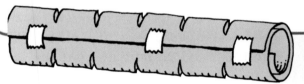

PLANTS AND WATER

Why do leaves give off water?

Materials
- pencil
- water
- marker
- scissors
- piece of thin cardboard
- leaf with a long stem
- modeling clay
- 2 clear plastic cups

Procedure

1. Carefully use the pencil to poke a hole in the center of the cardboard. Then push the leaf stem through the hole. Use the clay to close up the hole around the stem. Be careful not to pinch the stem.

2. Fill one cup about $\frac{2}{3}$ full with water. Mark the water line with the marker.

3. Snip off about 1 cm from the stem end. Place the cut stem into the water, resting the cardboard on the rim of the cup. Place the empty cup over the leaf. Set the cups in the sun.

4. After a few hours, observe both cups and the stem. Record your observations.

Draw Conclusions

What can you infer from your observations?

CHAPTER 2 Review and Test Preparation

Vocabulary Review

Use the terms below to complete the sentences. The page numbers in ( ) tell you where to look in the chapter if you need help.

classification (A38)
kingdom (A39)
monerans (A39)
protists (A39)
fungi (A39)
genus (A40)
species (A40)
vertebrate (A44)
mammal (A44)
reptile (A44)
amphibian (A44)
fish (A44)
bird (A45)
invertebrate (A45)
vascular plants (A50)
nonvascular plants (A52)

1. The largest group into which scientists classify living things is a ____.
2. An animal that does not have a backbone is an ____.
3. Plants are either ____ or ____, depending on the presence of tubes.
4. ____ and ____ are the smallest groups into which living things are classified.
5. An animal with a backbone is a ____.
6. The two kingdoms of microscopic living things are ____ and ____.
7. ____, ____, ____ are vertebrates.
8. ____ and ____ are also vertebrates.
9. Scientists use ____ to organize living things.
10. ____ have many cells with nuclei and absorb food from other living things.

Connect Concepts

Use the terms in the Word Bank to complete the concept map.

animals
classification
fungi
genus
kingdoms
monerans
plants
protists
species

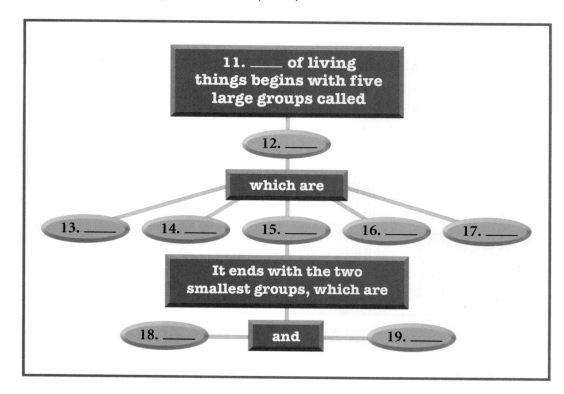

11. ____ of living things begins with five large groups called

12. ____

which are

13. ____ 14. ____ 15. ____ 16. ____ 17. ____

It ends with the two smallest groups, which are

18. ____ and 19. ____

A58

Check Understanding
Write the letter of the best choice.

20. Which of the following is a type of moneran?
    - A algae
    - B bacteria
    - C fish
    - D mushroom

21. Which characteristic makes vertebrates different from invertebrates?
    - F Vertebrates have a backbone.
    - G Vertebrates do not have a backbone.
    - H Vertebrates are monerans.
    - J Invertebrates have a backbone.

22. In a nonvascular plant, water travels —
    - A through the roots
    - B through the sapwood
    - C from cell to cell
    - D through the flowers

23. Which part of a vascular plant has tubes for carrying food?
    - F leaves
    - G heartwood
    - H bark
    - J ferns

24. Which of the following is NOT a kingdom?
    - A fungi
    - B animals
    - C plants
    - D vertebrates

25. Where do muscles attach to the skeletons of invertebrates?
    - F at flexible shell joints
    - G at the backbone
    - H where bones meet
    - J at movable bone joints

26. Which of the following vertebrates have hair and give milk for their young?
    - A reptiles
    - B mammals
    - C amphibians
    - D birds

27. What makes a moneran different from a protist?
    - F A moneran has no nucleus.
    - G A moneran has a backbone.
    - H A moneran has tubes.
    - J A moneran has jointed legs.

Critical Thinking

28. Why is it important that scientists share what they learn from their research?

29. A dog has a backbone and fur. To which kingdom and to which two smaller groups does it belong?

Process Skills Review

30. Which three items would you **classify** in one group? Explain your answer.

    shoelace, stop sign, button, zipper

31. Which would make the better **model** for showing how water is carried inside a tree? Explain your answer.

    a frozen-treat stick and paper

    a cardboard tube and a rubber hose

32. Think about your observations of the feet of ducks and chickens. Which animals would you **infer** are the better swimmers? Explain your answer.

Performance Assessment
Sorting Scheme

Work with a group to make rules for classifying items in your desks or in your classroom. Sort the items into several "kingdoms." Then sort the members of each kingdom into as many smaller groups as you can.

CHAPTER 3
Animal Growth and Heredity

Vocabulary Preview

chromosome
mitosis
asexual reproduction
sexual reproduction
meiosis
life cycle
direct development
metamorphosis
inherited trait
dominant trait
recessive trait
gene

Human children look like their parents. So do the young of many animals, such as this turtle. The young of many other organisms, however, do not look anything like their parents. Many organisms go through big changes as they grow and mature.

Fast Fact

It's sometimes hard to determine how long wild animals live, but we do know how long domesticated animals, like dogs and cats, and zoo animals live. Some can live as long as human beings.

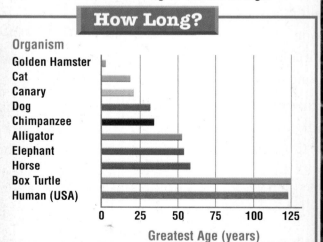

How Long?

Organism	Greatest Age (years)
Golden Hamster	
Cat	
Canary	
Dog	
Chimpanzee	
Alligator	
Elephant	
Horse	
Box Turtle	
Human (USA)	

Fast Fact

Many animals mature quickly. Mice, for example, start having young of their own when they are only four months old. By the time they are a year old, they may already be great-grandparents.

Fast Fact

About 3 billion of your body's cells die every minute. Thanks to a process called mitosis, about the same number of new cells are formed every minute.

Eastern box-turtle

LESSON 1

How Do Animals Grow and Reproduce?

In this lesson, you can . . .

 INVESTIGATE cell reproduction.

 LEARN ABOUT how organisms grow.

 LINK to math, writing, language arts, and technology.

 INVESTIGATE

Cell Reproduction

Activity Purpose Cells are the basic units of all living things. Cells make up bones, muscles, skin, and blood. They make up leaves, roots, stems, and flowers. Every part of an animal or a plant is made up of cells. As these cells grow and reproduce, the organism grows and develops. In this investigation you will **observe** how plant and animal cells reproduce.

Materials
- Microslide Viewer
- Microslide of plant mitosis
- Microslide of animal mitosis

Alternate Materials
- microscope
- prepared slides of plant cells dividing
- prepared slides of animal cells dividing

Activity Procedure

1. Insert the Plant Mitosis Microslide into the slot in the Microslide Viewer. Turn the focus knob until you can see the cells clearly. (Picture A)

2. **Observe** the plant cells dividing. **Record** what you observe in each stage of cell division. The descriptions on the microslide card may help you. Then draw pictures of what you see at each stage.

3. Now insert the Animal Mitosis Microslide into the slot in the Microslide Viewer. Again turn the focus knob until you can see the cells clearly.

◀ The cells that make up the tiger cub will reproduce rapidly as the cub grows into an adult.

Picture A

Picture B

4 **Observe** the animal cells dividing. **Record** what you observe in each stage of cell division. Again you may use the descriptions on the Microslide card to help you. Then draw pictures of what you see at each stage. (Picture B)

5 Now **compare** the stages of plant cell division with the stages of animal cell division. How are the stages alike? How are they different? **Record** your observations.

Draw Conclusions

1. What part of the cell changes as cell division occurs? What changes take place?

2. How many new cells does each dividing cell produce?

3. What similarities and differences did you **observe** between the dividing plant cells and the dividing animal cells?

4. **Scientists at Work** Scientists **observe** cells and ask questions based on their observations. What questions do you have about cell division, based on what you observed?

Investigate Further Now that you have observed photomicrographs of plant and animal cells dividing, use the materials on the *Alternate Materials* list to observe other cells dividing. See page R3 for tips on using a microscope.

> **Process Skill Tip**
>
> As you **observe** each slide, make a note of every detail—even if it doesn't seem important. Details will help you track exactly how a dividing cell changes from one stage to the next.

LEARN ABOUT

FIND OUT

- about growth and cell division
- about regeneration
- about mitosis and meiosis

VOCABULARY

chromosome
mitosis
asexual reproduction
sexual reproduction
meiosis

How Organisms Grow

Growth

You began life as a single cell. That one cell divided into two cells. The two cells divided into four cells, the four cells into eight cells, and so on. By the time you were born, your body was already made up of billions of cells.

It may be hard to observe, but you're probably a little taller than you were a month ago. In that short time, your bone cells have divided again and again to make more bone tissue. Muscle and skin cells have also been dividing. As cells throughout your body continue dividing, your body will continue growing. By the time you are an adult, your body will have more than 100 trillion cells.

You know that body systems are made up of organs, organs are made of tissues, and tissues are made up of cells. The cells that make up each tissue have specific functions as part of an organ. So when bone cells divide, for example, they must produce cells exactly like themselves. Nearly all body cells produce exact copies of themselves. Producing identical cells enables new cells to perform the same functions as older cells. This allows organs to continue functioning properly as they grow.

✓ **Why do bone cells make exact copies of themselves?**

DAY 1

A chick begins life as an egg—a single cell.

DAY 5

After 5 days the single cell has divided into many cells.

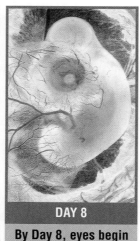

DAY 8

By Day 8, eyes begin to develop.

DAY 14

At 14 days the chick has developed tiny feathers.

DAY 21+

After 21 days the chick is fully developed.

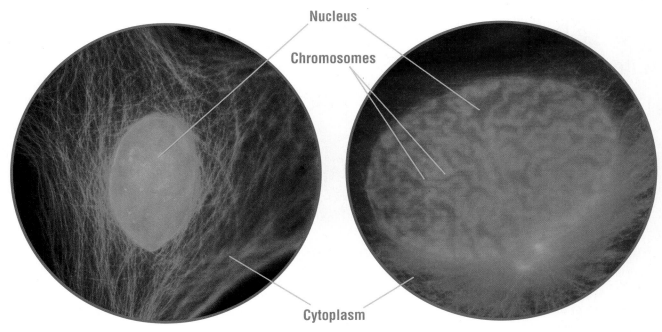

Before a cell divides, the chromosomes shorten, thicken, and become visible.

Cell Division

Every square centimeter of your skin has about 150,000 skin cells. The top layer of skin is damaged by contact with the world around you. Its cells are always dying. Fortunately, skin cells reproduce quickly and replace the dead cells. In fact, the surface layer of your skin will be replaced twice today. Like most cells, skin cells reproduce by dividing in two. But what makes cells divide?

You know that the nucleus controls everything a cell does. So it is the nucleus that "tells" a cell when to divide. Inside the nucleus are threadlike strands. Each strand is called a **chromosome** (KROH•muh•sohm). Chromosomes are made up of a chemical called DNA, which forms a chemical code. This code determines the shape and function of a cell. It also determines when a cell will divide.

Whenever a cell divides, each new cell must get an exact copy of the parent cell's chromosomes. Having an identical set of chromosomes gives each new skin cell, for example, the same DNA code as its parent cell. This ensures that it will look and act like older skin cells.

The process of cell division is called **mitosis** (my•TOH•sihs). The nucleus of a cell prepares for mitosis by making an exact copy of its chromosomes. After the chromosomes have been copied, but before mitosis begins, a cell has enough DNA for two cells. During mitosis each chromosome separates from its copy. The two groups of chromosomes pull apart. Then the cell membrane pinches in at the middle, forming two new cells. Each new cell has the same DNA—and therefore the same shape and function—as its parent cell.

✓ **What happens in the nucleus before a cell divides?**

THE INSIDE STORY

Mitosis

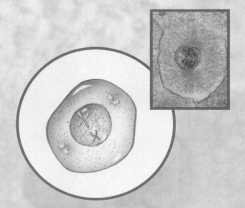
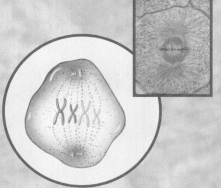

▲ Before mitosis the chromosomes (made up of DNA) make copies of themselves. Each chromosome and its copy are connected.

▲ As mitosis begins, the chromosomes grow thick and short. The nuclear membrane disappears.

▲ A network of thin tubes, called a spindle, forms. As the spindle gets longer, the chromosomes and their attached copies form a straight line in the middle of the cell.

This starfish escaped a hungry predator by dropping one of its arms. When regeneration is complete, the starfish will have a new arm. ▼

Regeneration

At one time or another, you have probably scraped your knee or cut your finger. If you cleaned the wound and kept it clean, after a few days all signs of the wound vanished. As soon as your body is wounded, it begins repairing itself.

Healing is a kind of *regeneration* (rih•jen•uh•RAY•shuhn), or tissue replacement. Skin cells divide, and regenerated skin grows over the cut. In humans, regeneration is limited mostly to healing wounds.

Plants and some animals, however, can regenerate major body parts. For example, if a large animal grabs a lizard by its tail, the tail may drop off. If this happens, the lizard's remaining tail cells begin rapid mitosis. Before long, the lizard has regenerated a new tail.

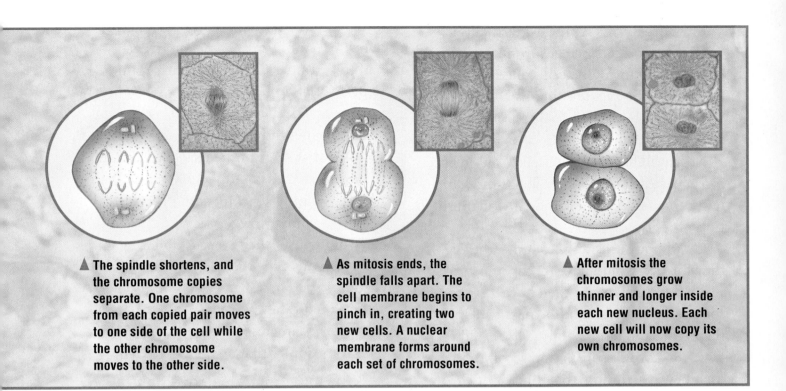

▲ The spindle shortens, and the chromosome copies separate. One chromosome from each copied pair moves to one side of the cell while the other chromosome moves to the other side.

▲ As mitosis ends, the spindle falls apart. The cell membrane begins to pinch in, creating two new cells. A nuclear membrane forms around each set of chromosomes.

▲ After mitosis the chromosomes grow thinner and longer inside each new nucleus. Each new cell will now copy its own chromosomes.

Planaria (pluh•NEHR•ee•uh), a type of freshwater flatworm, can regenerate most of its body. If a planarian is cut in two, each piece regenerates the missing parts through mitosis. And chopping a sponge into many pieces is never deadly. The pieces simply regenerate many new, complete sponges.

✓ **How does mitosis help heal a wound?**

Asexual Reproduction

Many one-celled organisms, such as bacteria and protozoa, reproduce by simple cell division, or *fission*. This type of reproduction—without the joining of male and female cells—is called **asexual reproduction**. In fission the parent produces offspring through the process of mitosis.

Yeast—a one-celled fungus—reproduces asexually by a process called *budding*. A tiny bud forms on the parent cell. Mitosis takes place within the parent cell, and a copy of the parent's chromosomes enters the growing bud. When the bud is fully grown, it separates from the parent cell.

Some plants also reproduce asexually. You will learn about asexual reproduction in plants in the next chapter.

✓ **How many parents are needed in asexual reproduction?**

A paramecium (par•uh•MEE•see•uhm) is a one-celled organism that lives in lakes and ponds. This paramecium is in the process of reproducing asexually. ▶

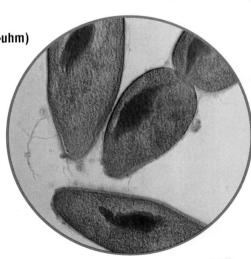

Sexual Reproduction

Most organisms reproduce sexually. In **sexual reproduction**, cells from two parents unite to form one cell, called a *zygote* (ZY•goht). The zygote contains chromosomes from both the female parent and the male parent.

Nearly all human body cells have 46 chromosomes each. If a body cell has more or fewer than 46 chromosomes, it won't function properly.

If two body cells were to unite, the zygote they formed would have 92 chromosomes. Every body cell would then have 92 chromosomes, instead of 46. In the next generation, the zygote and all the body cells would each have 184 chromosomes, and so on. With each new generation, the number of chromosomes would double.

This never happens, though, because the human body produces reproductive cells, which have only 23 chromosomes each—half the number of chromosomes found in body cells. **Meiosis** (my•OH•sis) is the process that reduces the number of chromosomes in reproductive cells.

In the first stage of meiosis, the cell copies its chromosomes and divides. Both of the new cells have 46 chromosomes. This process is similar to mitosis. In the second stage of meiosis, the two cells divide again. However, this time they do not copy their chromosomes first. Each of the four new cells is a reproductive cell, or *gamete* (GAM•eet). Gametes have half the number of chromosomes that are in a body cell.

✓ **Why must gametes have half the number of chromosomes that are in a body cell?**

Meiosis

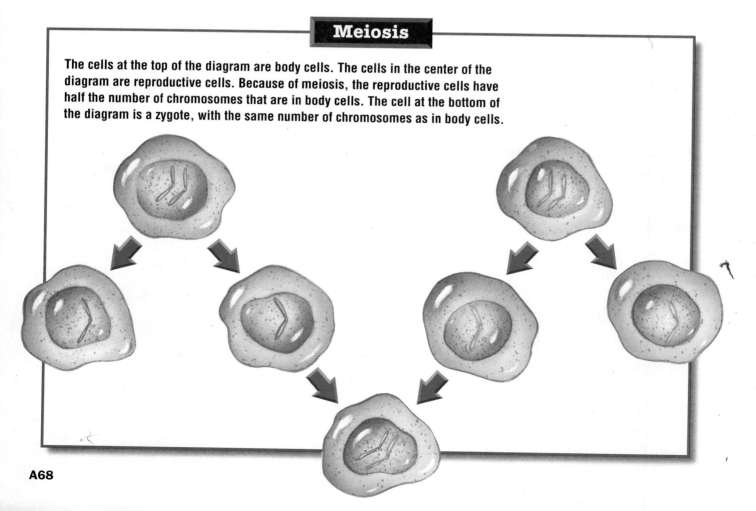

The cells at the top of the diagram are body cells. The cells in the center of the diagram are reproductive cells. Because of meiosis, the reproductive cells have half the number of chromosomes that are in body cells. The cell at the bottom of the diagram is a zygote, with the same number of chromosomes as in body cells.

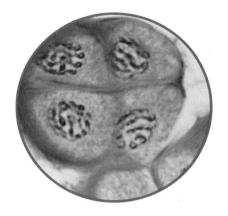

▲ Each of these gametes has half the number of chromosomes of a body cell.

Summary

Organisms grow when their cells divide. Body cells divide through the process of mitosis. Before mitosis, a cell makes copies of its chromosomes. After mitosis each of the new cells has an exact copy of the original cell's chromosomes. Reproductive cells have half the number of chromosomes that are in body cells. Meiosis reduces the number of chromosomes.

Review

1. What are chromosomes? Why are they important to cells?
2. How can some animals regenerate body parts?
3. Why must chromosomes be duplicated before mitosis?
4. What is asexual reproduction?
5. **Critical Thinking** Suppose an organism has 12 chromosomes in its body cells. How many chromosomes do its reproductive cells and zygotes have?
6. **Test Prep** During mitosis, what part of the cell pinches in to make two new cells?
   A the nucleus
   B the chromosomes
   C the DNA
   D the cell membrane

LINKS

MATH LINK

Use Divisibility Rules The list of organisms below shows the numbers of chromosomes in their body cells. Determine how many chromosomes each organism's reproductive cells have.

alligator, 32   housefly, 12
earthworm, 36   chimpanzee, 48
pigeon, 80   lettuce, 18

WRITING LINK

Informative Writing—Explanation Write an article for your school newspaper explaining the importance of meiosis. Briefly describe meiosis, and explain how it enables an organism to keep the same number of chromosomes in its body cells after sexual reproduction.

LANGUAGE ARTS LINK

Sequencing Mitosis Cut a sheet of paper into 10 pieces. On each piece, write something that happens during mitosis—for example, *spindle forms* or *nuclear membrane disappears*. Exchange pieces of paper with a classmate. Challenge each other to put the pieces in order.

TECHNOLOGY LINK

Learn more about cell division and reproduction by visiting this Internet site.
www.scilinks.org/harcourt

LESSON 2

What Is a Life Cycle?

In this lesson, you can . . .

INVESTIGATE the stages of a mealworm's life.

LEARN ABOUT the stages of life.

LINK to math, writing, art, and technology.

◀ This butterfly is about to begin the adult stage of its life cycle.

The Stages of a Mealworm's Life

Activity Purpose Many animals change as they grow and develop. The way an animal behaves and looks when it is young may change by the time it is an adult. In this investigation you will **observe** how a mealworm changes as it grows.

Materials
- mealworm culture
- paper plate
- hand lens

Activity Procedure

1. Your teacher will give you a mealworm from the mealworm culture.

2. Put the mealworm on the paper plate. Using the hand lens, **observe** the mealworm closely. Draw what you see. (Picture A)

3. Label these parts on your drawing: head, segment, antenna, outer shell, claw, mouth, and leg.

4. **Observe** the mealworm's movements. Does it move straight forward or from side to side? Does it move quickly or slowly? **Record** your observations.

5. Now your teacher will give you a beetle from the mealworm culture. Repeat Steps 2–4 with the beetle.

Picture A

Picture B

6 Finally, **observe** the mealworm culture. Try to find evidence of other stages of a mealworm's life, such as eggs and pupa cases. Draw pictures of what you find.

7 **Compare** your drawings of the eggs and pupa cases with your drawings of the mealworm and the beetle. Form a **hypothesis** about the order of these life stages. Then list the ways in which a mealworm changes as it grows. (Picture B)

Draw Conclusions

1. How are the mealworm and the beetle similar?
2. How is the beetle different from the mealworm?
3. **Scientists at Work** Scientists often **observe** an organism and then **hypothesize** about their observations. What observations enabled you to form your hypothesis about the order of the life stages?

Investigate Further **Plan and conduct an experiment** to test your hypothesis about the life stages of a mealworm. Decide what equipment or technology you will need to use to test your hypothesis. Then use the equipment or technology in your experiment.

Process Skill Tip

As you **observe** stages in the life of an organism, you may find it helpful to **hypothesize** about the order in which the stages occur.

LEARN ABOUT

FIND OUT
- what life cycles are
- how animals change as they grow
- how complete metamorphosis differs from incomplete metamorphosis

VOCABULARY

life cycle
direct development
metamorphosis

The Stages of Life

Life Cycles

Most organisms grow and mature through several distinct stages of life. These stages make up the organism's **life cycle**. All life cycles begin with a young organism struggling to survive in the world. Some organisms are born alive. Others develop in eggs and then hatch. Still others sprout from spores or seeds.

During each stage of life, a person changes physically. ▶

Depending on its type, a young organism spends anywhere from a few minutes to many years growing and developing. When an organism reaches its final form and size, it is an adult. During the adult stage, an organism is able to reproduce. Some organisms reproduce during their entire adult lives. Others stop reproducing as they get older.

Many organisms change a lot as they mature. But the young of some animals are identical to the adults, except in size. These organisms grow larger, but they keep the same body features, such as shape, all their lives. This kind of growth is called **direct development**. Spiders and earthworms have direct development. A young earthworm looks just like an adult earthworm, except smaller. What other animals can you think of that show this type of development?

These young scorpions are identical to the adult carrying them, except in size and color. ▼

✓ **What is a life cycle?**

A72

Incomplete Metamorphosis

Some animals, especially insects, have one kind of body when they are young and a very different kind of body when they are adults. The changes in the shape or characteristics of an organism's body as it grows and matures are called **metamorphosis** (met•uh•MOR•fuh•sis).

The life cycles of many insects, including beetles, butterflies, and grasshoppers, include metamorphosis. Some insects, such as cockroaches and grasshoppers, go through *incomplete metamorphosis*. These insects have only three stages of development: egg, nymph (NIMF), and adult. At each stage the insect looks different from the way it looks at another stage.

When a cockroach nymph first hatches, it doesn't have wings. After a period of growth, tiny wing buds appear on its body. As the nymph continues growing, wings develop from the wing buds. It takes about three months for a young cockroach to become an adult with fully formed wings.

During late spring and early summer, you may come across tiny, wingless hopping insects that look like grasshoppers. They are, in fact, grasshopper nymphs.

Cockroaches, grasshoppers, and other insects that go through incomplete metamorphosis must shed their outer skeletons as they grow. This process is called *molting*. Each time a grasshopper molts, a larger skeleton forms around its body. This gives the insect room to grow for a while.

✓ **Name the stages of incomplete metamorphosis.**

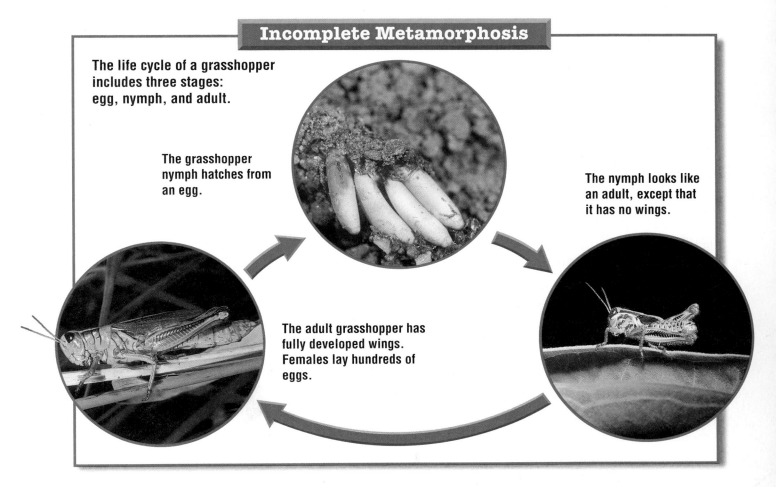

Incomplete Metamorphosis

The life cycle of a grasshopper includes three stages: egg, nymph, and adult.

The grasshopper nymph hatches from an egg.

The nymph looks like an adult, except that it has no wings.

The adult grasshopper has fully developed wings. Females lay hundreds of eggs.

Complete Metamorphosis

The life cycle of a beetle includes four stages: egg, larva, pupa, and adult.

◀ An adult beetle lays hundreds of eggs.

When a pupa has become an adult, a mature beetle emerges. ▼

The pupa changes a larva into an adult. ▶

An egg develops into a larva. ▲ The larva, sometimes called a grub, eats plant roots, stems, and leaves.

Complete Metamorphosis

Have you ever been startled by a fuzzy, many-legged caterpillar creeping along a tree trunk? It's hard to imagine, but that same creature may one day be a moth or beautiful butterfly. Animals whose bodies change dramatically during their life cycles go through *complete metamorphosis*. A beetle is another example of an animal that goes through complete metamorphosis.

An insect's life cycle that includes complete metamorphosis has four distinct stages. During the first stage, the insect develops inside an egg. When the insect hatches from the egg, it enters the second stage of development. It is then called a *larva* (LAR•vuh). A larva lacks wings and looks very different from an adult. It spends nearly all its time eating and storing energy for the next stage of its life cycle.

During the third stage of complete metamorphosis, insects neither eat nor move. At this stage, an insect is called a *pupa* (PYOO•puh). The pupa's body breaks down many of the organs and tissues it had as a larva. Most of the pupa's energy goes into developing an adult body. When the stage is complete, the pupa no longer exists and the adult insect emerges. Like all adult insects, the female lays eggs, the beginning of a new life cycle.

✓ **What are the stages of complete metamorphosis in insects?**

▲ This palm beetle lives in the Amazon rain forest.

Summary

All organisms have life cycles. Most organisms begin as young and develop into adults. The life cycles of some organisms show direct development. These organisms change only in size as they mature. Other life cycles include incomplete metamorphosis, with three stages of development. A life cycle that includes complete metamorphosis has four stages of development.

Review

1. What is the last stage in the life cycle of a beetle?
2. In what kind of development do the young look like miniature adults?
3. Why does a grasshopper nymph molt?
4. **Critical Thinking** Why doesn't a developing pupa move around?
5. **Test Prep** During the third stage of complete metamorphosis, the larva becomes—

    A a cocoon
    B a pupa
    C an egg
    D an adult

LINKS

MATH LINK

Solve Problems Monarch butterflies and June bugs go through complete metamorphosis. A monarch caterpillar (the larva) eats milkweed leaves for about a month before becoming a pupa. A June bug larva feeds underground for about three years. How many times as long as a monarch caterpillar eats does a June bug larva eat?

WRITING LINK

Expressive Writing—Poem Choose an animal, such as a frog, and write a poem for your class about its life cycle.

ART LINK

Mobile Research the complete metamorphosis of the monarch butterfly. Then make a hanging mobile illustrating the four stages of the monarch's life cycle. Share your mobile with your class.

TECHNOLOGY LINK

Learn more about complete metamorphosis of butterflies by visiting the Smithsonian Institution Internet site.
www.si.edu/harcourt/science

Smithsonian Institution®

LESSON 3

Why Are Offspring Like Their Parents?

In this lesson, you can . . .

 INVESTIGATE inherited characteristics.

 LEARN ABOUT inherited traits.

 LINK to math, writing, social studies, and technology.

◀ This child and parent share obvious traits.

Inherited Characteristics

Activity Purpose What color are your hair and eyes? How long are your eyelashes? Are you right-handed or left-handed? These are just a few of the characteristics you have gotten from your parents. You may have some traits in common with your classmates. In this investigation you will **observe** three inherited traits. Then you will **use numbers** to see how common those traits are among your classmates.

Materials
- mirror

Activity Procedure

1. Make a chart like the one on the next page.

2. **Tongue Rolling** Use the mirror to **observe** what you are doing. Stick out your tongue, and try to roll its edges up toward the center. **Record** your results in the chart. (Picture A)

3. **Ear Lobes** Use the mirror to **observe** the shape of your ear lobes. Are they attached to your cheek, or do they hang free? **Record** your results in the chart.

4. **Folded Hands** Clasp your hands in front of you. **Observe** which of your thumbs falls naturally on top. **Record** your results in the chart. (Picture B)

Picture A

Picture B

5. Your teacher will now ask students to report the results of their observations. Tally the results in the chart as students report them. Total the number of students for each characteristic. Then **calculate** what fraction of the class has each characteristic.

Characteristic	Results (Circle one.)		Class Totals
Tongue rolling	Yes	No	
Ear lobes	Attached	Free	
Folded hands	Left	Right	

Draw Conclusions

1. **Infer** whether a person could learn tongue rolling. Explain.
2. What other inherited characteristics could you have **observed**?
3. **Scientists at Work** Scientists often **use numbers** to summarize the data they collect. Which trait in each pair occurred most often in your class?

Investigate Further Do your class results suggest how often these traits occur in other people? Choose one or two of these characteristics. **Hypothesize** whether the results will be the same for another group. Then ask some of your friends, neighbors, and family members to participate in this activity, and **collect data. Draw conclusions,** and share them with your class.

> **Process Skill Tip**
>
> When you **use numbers**, you can summarize the data you collect.

Inherited Traits

From Parent to Offspring

FIND OUT
- how characteristics are inherited
- what genes are

VOCABULARY
inherited trait
dominant trait
recessive trait
gene

You may have friends who look a lot like one of their parents. They might share the same smile, eyes, type of hair, or skin color. Many of the characteristics of organisms are passed to their offspring. A characteristic that is passed from parent to offspring is an **inherited trait**. In many animals, hair color or fur color is an inherited trait. So is eye color.

Sometimes an offspring has a trait that it doesn't seem to have inherited from its parents. Two brown-haired human parents might have a child with blond hair. In the picture below, observe that the mother cat is orange. All but one of her kittens are orange, too. The gray kitten must have inherited its fur color from its parents. But how could orange parents produce a gray kitten?

Some behaviors are also inherited. Most dogs, for example, can swim without having been taught. But this is not an inherited behavior among humans.

✓ **What is an inherited trait?**

Sometimes a trait doesn't seem to have been inherited. ▼

How Characteristics Are Inherited

◀ Gregor Mendel

Gregor Mendel, a monk and a scientist, spent much of his time working with plants in a monastery garden. He observed that some pea plants were tall while others were short. He also observed that some plants produced green peas and others produced yellow peas. He knew the traits were inherited, but he didn't understand how.

In 1857 Mendel began experimenting by breeding different pea plants. This is called *crossbreeding*. He kept detailed records of his results, which were published in 1865.

Mendel first bred a tall pea plant with a short pea plant. All the offspring, the first generation, were tall plants. He then bred two of these tall plants. In the second generation, three-fourths of the offspring were tall and one-fourth were short.

Mendel hypothesized that every trait is controlled by a pair of *factors*. For each trait, an offspring inherits one factor from each parent. The way the factors combine determines which trait appears in the offspring.

Mendel further hypothesized that the first-generation pea plants must have a hidden factor for shortness. Why? In the second generation, one-fourth of the offspring were short.

In peas, tallness is a strong trait, or **dominant trait**. Shortness is a weak trait, or **recessive trait**. Factors for both dominant and recessive traits may occur in an organism's chromosomes. However, recessive traits can be seen only if both parents pass the factor for it to the offspring. If only one parent pea plant, for example, passes a factor for shortness, the factor remains hidden. Then none of the offspring are short.

✓ **How did Mendel know that the tall pea plants in the first generation had a hidden factor for shortness?**

Inherited Factors

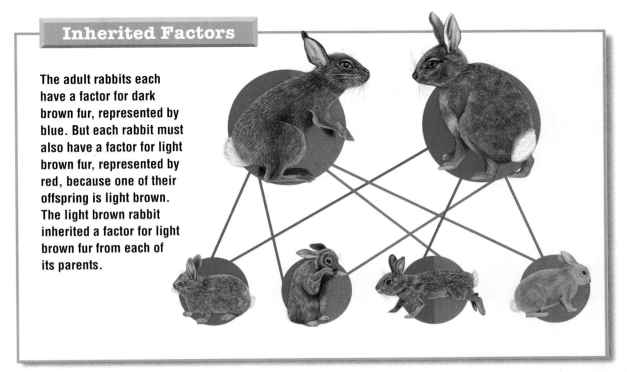

The adult rabbits each have a factor for dark brown fur, represented by blue. But each rabbit must also have a factor for light brown fur, represented by red, because one of their offspring is light brown. The light brown rabbit inherited a factor for light brown fur from each of its parents.

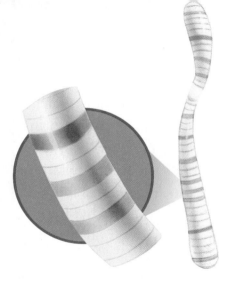
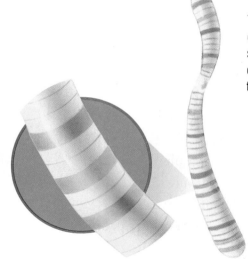

The gene for fur color occurs in the same spot on both chromosomes of the pair.

Genes

Today Mendel's factors are called genes. **Genes** contain the DNA codes for all the traits an organism inherits. Genes are on chromosomes. The genes for a particular trait always occupy the same place on a chromosome.

In wild rabbits, dark brown fur is a dominant trait. Light brown fur is a recessive trait. If a rabbit inherits even one gene for dark brown fur, it will have dark brown fur. Mendel's hidden factors were genes for recessive traits. Offspring must inherit two genes for recessive traits—one from each parent—for a recessive trait to be seen.

You can use a chart to calculate the chances that an offspring will inherit a particular combination of genes. The charts below show the chances that the offspring of two rabbits will inherit a particular fur

Dominant Traits

The chart at the left shows that all the offspring have dark brown fur. So the dark brown rabbit had two genes for dark brown fur. The chart at the right shows that half the offspring will have dark brown fur and half will have light brown fur. So the dark brown rabbit had one gene for dark brown fur and one gene for light brown fur. A trait with two identical genes is called *purebred*. The same trait with two different genes is called *hybrid*.

The dark brown rabbit may be purebred, having two genes for dark brown fur, or it may be hybrid, having one gene for dark brown fur and one gene for light brown fur. The light brown rabbit must be purebred and have two genes for light brown fur. ▼

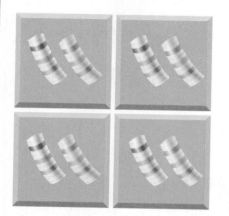

color. Dark brown fur is a dominant trait. The gene for dark brown fur is highlighted in purple. Any box with a purple-highlighted gene in it stands for an offspring with dark brown fur. Light brown fur is a recessive trait. The gene for light brown fur is highlighted in orange. Only a box with two orange-highlighted genes stands for an offspring with light brown fur.

✓ **Why must baby rabbits have two genes for light brown fur if they are to have light brown fur?**

Summary

Many traits are inherited by offspring from their parents. Gregor Mendel's experiments with peas led to the discovery of the factors of inheritance. Genes—Mendel's "factors"—are on chromosomes. Each trait is controlled by a pair of genes. Gene combinations determine if traits are seen (dominant traits) or hidden (recessive traits).

Review

1. What are characteristics called that an offspring inherits from its parents?
2. What did Mendel hypothesize about inherited traits?
3. What is a dominant trait?
4. **Critical Thinking** Will an offspring show a recessive trait if only one of its parents has the gene for that trait? Explain.
5. **Test Prep** Mendel's "factors" are now known as —
   A genes
   B chromosomes
   C DNA
   D traits

LINKS

MATH LINK

Find Probability If you toss a coin 100 times, you might expect it to land heads up 50 times and tails up 50 times. That is, the chance, or probability, of its landing heads up is the same as the probability of its landing tails up. What is the probability that two coins, flipped at the same time, will both land heads up? Make a table with four columns: *Heads/Heads, Heads/Tails, Tails/Heads, Tails/Tails.* Then get two coins from your teacher. Flip them together 100 times. Record your results in the table. What is the probability that two coins will land heads up?

WRITING LINK

Expressive Writing—Song Lyrics Write a song for a pet about one of his or her best traits, such as the color of his or her eyes or fur.

SOCIAL STUDIES LINK

Research Obtain a book about James Watson and Francis Crick, the scientists who helped discover the structure of DNA. Write a paragraph about what they did.

TECHNOLOGY LINK

Learn more about inherited traits by viewing *Better Turkeys* on the **Harcourt Science Newsroom Video.**

SCIENCE AND TECHNOLOGY
BIONIC DOG

As dogs grow older, they sometimes develop a painful disorder called hip dysplasia (dihs·PLAY·zhuh). Now veterinarians are able to replace one or both damaged hips with artificial parts.

Why Treat Hip Dysplasia?

Hip dysplasia is an abnormal development and growth of the hip joint, leading to arthritis. It causes pain whenever a dog stands, walks, runs, or jumps. In the early stages of the disorder, medicine can help, but when the pain becomes severe, medication is not enough. Dogs who were once very active no longer want to move. As the disorder progresses, the dog feels pain all the time, even when lying down. A dog may suffer so much that it has to be "put to sleep."

Total Hip Replacement (THR)

A total hip replacement, or THR, can save many dogs with hip dysplasia. In THR the damaged hip is replaced with a steel and plastic joint, often called an implant. Most dogs can walk immediately after surgery. After a recovery period of about two months, they can run, jump, and play as if they had a normal hip. Over 95 percent of all THR surgeries performed on dogs are successful, and the new hip lasts for the rest of the dog's life.

Large dogs are more likely than small dogs to have hip dysplasia.

How Does THR Work?

Any dog that is at least 9 months old and weighs at least 30 pounds can be considered for THR. A few hours before the surgery, the hair on the dog's leg is shaved and the skin bathed with medicine to prevent infection. During the surgery, which takes 90–120 minutes, the dog is given an anesthetic.

The artificial hip the dog receives consists of three parts. The femoral stem attaches to the femur in the dog's leg. The femoral head replaces the ball of the hip's ball-and-socket joint. A part called the acetabular cup replaces the original socket of the hip joint.

The implants are cemented to healthy bones with surgical cement. Hip implants for dogs are made from the same materials and implanted the same way as those for human hip replacements.

Instead of a THR, a dog can have a femoral head osteotomy, which is removal of the top of the femur of the diseased hip. However, this leaves the dog with less movement than a THR. For a very young dog with hip dysplasia, surgery is done to save as much of the damaged hip as possible.

THINK ABOUT IT

1. Why do you think hip dysplasia usually occurs in bigger dogs rather than smaller dogs?
2. Why do you think a young pup is not given a THR?

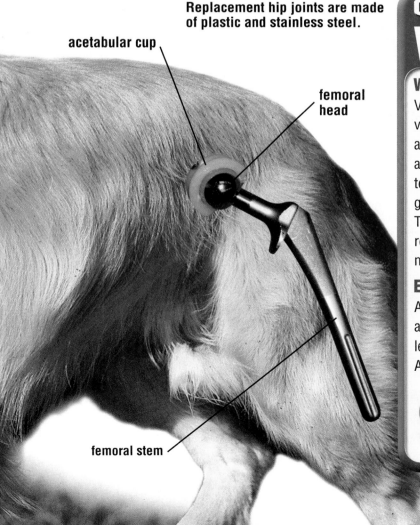

Replacement hip joints are made of plastic and stainless steel.

- acetabular cup
- femoral head
- femoral stem

CAREERS
VETERINARY TECHNICIAN

What They Do
Veterinary technicians, or vet techs, assist veterinarians in private practice and in animal hospitals. Some vet techs assist veterinary surgeons who perform THRs. They may also work as research assistants in pharmaceutical companies.

Education and Training
A person wanting to become a vet tech should attend a two-year or four-year college accredited by the American Veterinary Medical Association and major in animal science technology.

WEB LINK
For Science and Technology updates, visit the Harcourt Internet site.
www.harcourtschool.com

PEOPLE IN SCIENCE

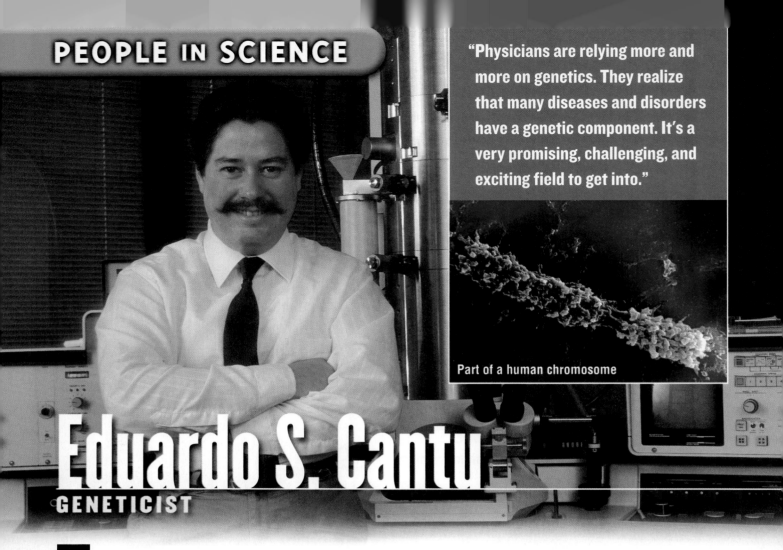

"Physicians are relying more and more on genetics. They realize that many diseases and disorders have a genetic component. It's a very promising, challenging, and exciting field to get into."

Part of a human chromosome

Eduardo S. Cantu
GENETICIST

Every human baby is born with about 30,000 different genes—and the possibility of more than 3000 genetic disorders. Most of these disorders are very rare, but geneticist Eduardo S. Cantu says that 50 to 100 of them could be considered common. Dr. Cantu, a scientist at the Medical University of South Carolina, studies ways in which genes and chromosomes affect a person's health.

As director of the Medical University's cytogenetic section, Dr. Cantu and his staff run tests that help physicians diagnose genetic disorders. Most of the tests they perform are used to diagnose genetic disorders before birth. The physician collects some of the fluid that surrounds a developing fetus, or unborn child. Then Dr. Cantu and his staff examine the fluid, which contains cells from the fetus, under a microscope. The genes in the cells provide a great deal of information about the unborn child, including whether it is likely to develop certain kinds of cancers and other diseases.

Dr. Cantu, who was born and raised in Texas, began his science education at Pan American University. His interest in genetics began when he was a graduate student. He had intended to study birds, but he found a genetics course so fascinating that he decided to change his course of study. It is a decision that Dr. Cantu is very glad he made.

THINK ABOUT IT

1. What is one kind of information that can be gained by studying the genes in the cells of a fetus?
2. Why do you think genetic research is a growing field of study?

ACTIVITIES FOR HOME OR SCHOOL

DNA

How does DNA fit into a cell?

Materials
- sewing thread
- meterstick
- gelatin capsule

Procedure
1. Measure 10 m of thread. The thread stands for the DNA in a cell.
2. Open the gelatin capsule. The gelatin capsule stands for the nucleus of a cell.
3. Put all the thread into the capsule in whatever way you can. Then close the capsule.

Draw Conclusions
How did you get all the thread into the gelatin capsule? How does so much DNA fit into the nucleus of a cell?

CHROMOSOMES

How do chromosome pairs separate?

Materials
- embroidery thread
- meterstick

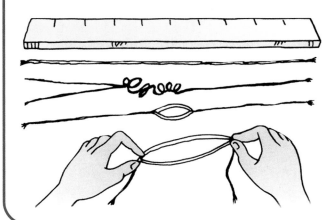

Procedure
1. Measure 1 m of embroidery thread. The thread represents a chromosome.
2. Starting at one end, separate the thread into two pieces.
3. Again measure 1 m of embroidery thread.
4. Now separate the thread into two pieces, starting in the middle.

Draw Conclusions
Was it easier to separate the thread into two pieces by starting from one end or from the middle? During mitosis, where do chromosome pairs separate?

CHAPTER 3 Review and Test Preparation

Vocabulary Review

Use the terms below to complete the sentences. The page numbers in ( ) tell you where to look in the chapter if you need help.

chromosomes (A65)
mitosis (A65)
asexual reproduction (A67)
sexual reproduction (A68)
meiosis (A68)
life cycle (A72)
direct development (A72)
metamorphosis (A73)
inherited trait (A78)
dominant trait (A79)
recessive trait (A79)
gene (A80)

1. The term ____ applies to an organism whose body shape, form, or characteristics change during its life cycle.

2. In humans, hair color is an example of an ____.

3. The process that reduces the number of chromosomes in certain cells is called ____.

4. An organism in the adult stage of its ____ is able to reproduce.

5. The offspring of some animals are just like the adults, except for size. This is an example of ____.

6. During ____, duplicated ____ in the cell's nucleus separate, and the cell pinches into two cells.

7. In order for a ____ to be seen, an offspring must receive a factor for the trait from each parent. In order for a ____ to be seen, an offspring needs to receive a factor from only one parent.

8. A ____ carries the DNA code for a particular trait.

9. Reproduction without the joining of cells from the female parent and the male parent is called ____. Reproduction in which cells from two parents unite to form a zygote is called ____.

Connect Concepts

Choose terms from the Word Bank that complete the concept map.

mitosis   recessive
function   nucleus

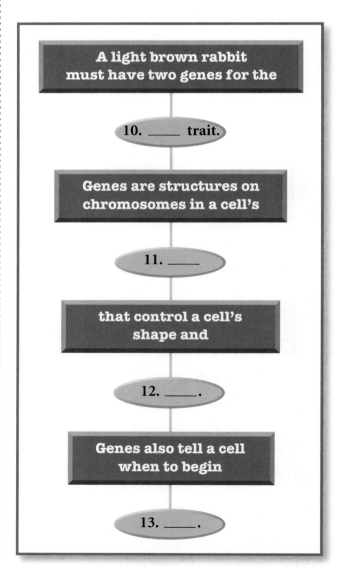

A light brown rabbit must have two genes for the

10. ____ trait.

Genes are structures on chromosomes in a cell's

11. ____

that control a cell's shape and

12. ____.

Genes also tell a cell when to begin

13. ____.

Check Understanding
Write the letter of the best choice.

14. In order for a new cell to function just like its parent cell, it must receive an exact copy of its parent cell's —
    A chromosomes   C cell wall
    B gametes   D recessive traits

15. Your bones and muscles grow larger as a result of —
    F meiosis   H direct development
    G metamorphosis
    J mitosis

16. What does **NOT** happen during mitosis?
    A the chromosomes duplicate
    B a spindle fiber forms
    C the number of chromosomes is halved
    D the nuclear membrane disappears

17. All the traits that parents pass to their offspring are called —
    F recessive traits   H physical traits
    G dominant traits   J inherited traits

18. If an offspring receives two genes, one from each parent, for a particular recessive trait, that trait will —
    A not be seen   C be seen halfway
    B be seen   D none of the above

19. Which sequence illustrates incomplete metamorphosis?
    F egg, larva, pupa, butterfly
    G kitten, cat
    H egg, nymph, adult
    J young raccoon, adult raccoon

20. When meiosis occurs, the number of chromosomes —
    A is doubled   C stays the same
    B is halved   D is tripled

Critical Thinking

21. A dark brown rabbit mates with a light brown rabbit. Out of four offspring, how many are likely to be dark brown?

22. Mitosis is necessary for the individual, but meiosis is necessary for the species. Explain this statement.

Process Skills Review

23. Choose an item in your classroom. It may be living or nonliving. **Observe** the item using as many of your senses as possible. Make at least seven observations about the item's characteristics.

24. **Compare** the life cycle of a grasshopper with that of a butterfly. Which insect's life cycle shows complete metamorphosis?

25. Describe the expected results of a cross between a dark brown rabbit and a light brown rabbit. **Use numbers** to describe the results in the second generation.

Performance Assessment
Charting Traits

Consider two inherited traits, such as tongue rolling and free or attached ear lobes. Choose symbols to represent genes for the dominant traits and genes for the recessive traits. For each trait, make charts showing the possible offspring when one parent has the dominant traits and the other parent has the recessive traits.

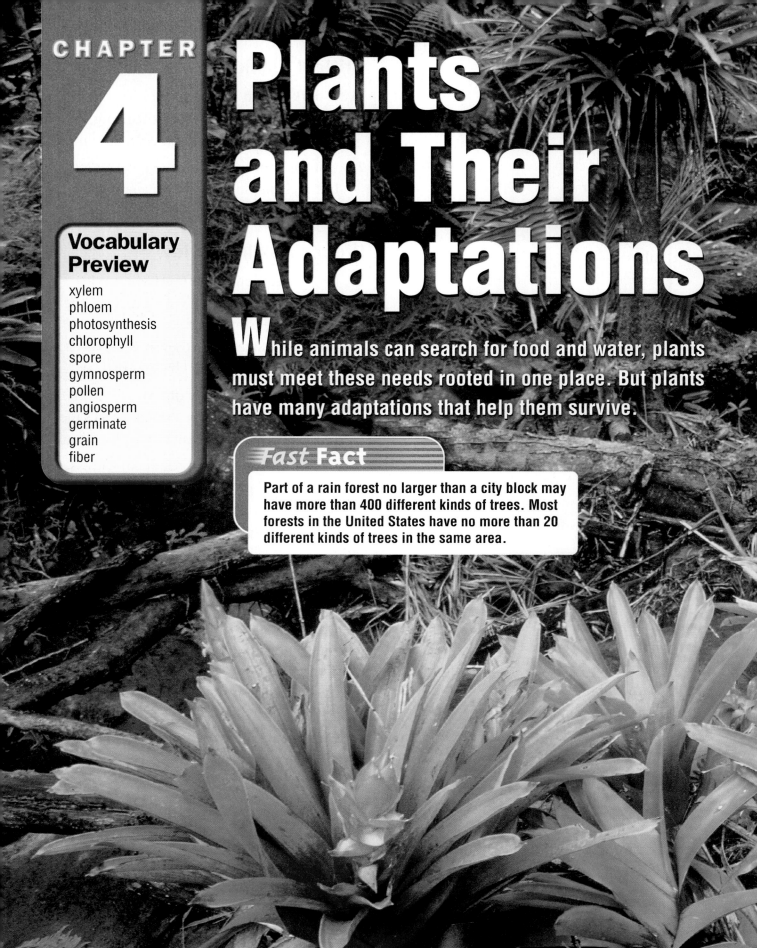

CHAPTER 4

Plants and Their Adaptations

Vocabulary Preview

xylem
phloem
photosynthesis
chlorophyll
spore
gymnosperm
pollen
angiosperm
germinate
grain
fiber

While animals can search for food and water, plants must meet these needs rooted in one place. But plants have many adaptations that help them survive.

Fast Fact

Part of a rain forest no larger than a city block may have more than 400 different kinds of trees. Most forests in the United States have no more than 20 different kinds of trees in the same area.

Fast Fact

Have you ever heard the expression "as American as apple pie"? It may surprise you to learn that the apple is not a native American plant and that apple pie was invented in Europe.

Food Origins

Food	Probable Origin
Apple	Southeast Europe
Banana	Southeast Asia
Oats	Western Europe
Pineapple	Latin America
Potato	Peru
Rice	India
Tomato	Mexico

Fast Fact

Buffalo grass has a thick network of roots. These roots and the soil they hold are called *sod*. The roots hold the soil so well that many settlers on the American plains used sod to build their homes.

LESSON 1

What Are the Functions of Roots, Stems, and Leaves?

In this lesson, you can . . .

 INVESTIGATE the parts of plants.

 LEARN ABOUT plant parts and their functions.

 LINK to math, writing, social studies, and technology.

The silver sword grows only in Hawai'i.

The Parts of a Vascular Plant

Activity Purpose Towering oak trees and potted geraniums have many parts in common. Vascular plants have roots, stems, and leaves. The sizes and shapes of these parts differ, but each part has a function that helps the plant live. In this investigation you will **observe** a plant and identify its parts.

Materials
- potted plant
- hand lens
- ruler
- newspaper
- plastic knife

Activity Procedure

1. Make a drawing of the plant. List all the parts of the plant that you can name.

2. **Observe** the leaves. What colors are they? Use the ruler to measure the length and width of the leaves. Are they all the same shape and size? Are they wide or narrow? Are they long or short? Do they grow singly or in pairs? Observe them more closely with the hand lens. What more can you say about them? Identify and label the leaves in your drawing. (Picture A)

3. **Observe** the stem. Does it bend? Does it have branches? Identify and label the stem in your drawing.

4. Hold the pot upside down over the newspaper. Tap the pot gently until the plant and the soil come out. If the plant won't come out, run the plastic knife around between the soil and the inside of the pot. (Picture B)

Picture A

Picture B

5. Shake the soil from the roots until you can see them clearly. **Observe** the roots. Is there a single root, or are there many small roots? What shape are the roots? Use the ruler to **measure** the length of the roots. Are they thick or thin? Long or short? Use the hand lens to observe them more closely. What more can you say about them? Identify and label the roots in your drawing.

6. Put the soil and the plant back into the pot. Water the plant lightly to help it recover from being out of the pot.

Draw Conclusions

1. What are the parts of the plant you **observed**?

2. **Compare** the plant parts you identified with the parts of a large tree. How are they the same? How are they different?

3. **Scientists at Work** Scientists learn by making observations. What did you **observe** about each part of the plant?

Investigate Further What questions about plant parts could you answer if you had other measuring tools? **Form a hypothesis** about the functions of plant parts. Then **plan and conduct an experiment** to test your hypothesis.

Process Skill Tip

When you **observe** something, you should use as many of your senses as you can. Don't just look at the plant. Touch it to see what it feels like and how thick or strong or dry it is. Smell its leaves and roots.

A91

LEARN ABOUT

What Vascular Plant Parts Do

FIND OUT

- how vascular plants grow in different environments
- what roots, stems, and leaves do

VOCABULARY

xylem
phloem
photosynthesis
chlorophyll

Common Parts

There are more than half a million types of vascular plants on Earth. They range from tiny desert plants, smaller than a pencil eraser, to giant redwood trees, taller than a 25-story building. No matter how different they appear, vascular plants have three parts in common—roots, stems, and leaves.

These parts make it possible for vascular plants to live and grow almost everywhere. Vascular plants are found in dry deserts, wet jungles, and cold Arctic plains. Vascular plants are able to live in different environments because their roots, stems, and leaves are adapted to the environments in which they live.

✓ **What parts are common to vascular plants?**

◀ Some plants are able to grow in unfavorable locations.

These crocuses bloom in low temperatures that would kill many other plants. ▼

Some tall trees have roots that spread out as far as their branches. They help keep the trees from falling over.

A thick mat of fibrous roots allows grasses to take in large amounts of water from the soil. ▶

Dandelions have long taproots. If you try to pull a dandelion out of the ground, part of the root may remain. The dandelion will grow back from the part that is left in the ground. ▶

Roots

The roots of many trees spread as far or farther from the trunk as their branches do. Most plant roots act as anchors. They keep the plant from falling over or blowing away in the wind. Roots also take in water and nutrients from the soil through tiny parts called *root hairs*. Some roots also store food for the plant.

Different types of roots are adapted to different environments. Some small desert plants have roots that spread far from the plant but grow close to the surface. These roots are able to take in large amounts of the little rain that falls in the desert.

Forest trees don't need the spreading roots of desert plants because there is more water in the forest soil. These trees need deep roots to anchor them. Some tree roots, called *prop roots*, begin above the ground. These help keep trees that grow in loose, wet soil from being blown over by the wind.

Many plants have *fibrous* (FY•bruhs) *roots*, which look a little like tree branches. The fibrous roots of some grasses form a thick and tangled mat just under the surface of the soil. Fibrous roots help prevent soil erosion by wind and water because they anchor the soil as well as the plant.

Some plants have a single, thick root that grows straight down. These roots, called *taproots,* can reach water that is deep in the ground. Some taproots store food for the plant as well.

In tropical rain forests there are plants that grow on branches high in the trees. Their roots attach themselves to the trees and take water directly from the moist air.

✓ **What do plant roots do?**

Storage Roots

Some plants store extra food and water to help them survive brief changes in their environments. Most plants cannot make food in the winter. In dry periods they may not be able to get all the water they need from the soil.

In good weather plants produce more food than they need and take in extra water. Some plants store extra food and water in their roots. Others store it in their stems.

Some plants store so much extra food in their roots that people grow them for their own food. You've probably eaten several kinds of storage roots. Beets, carrots, sweet potatoes, and turnips are called root vegetables. Because much of the food they store is in the form of sugar and starch, many root vegetables have a sweet taste.

✓ **What do some roots store?**

Stems

Stems do several things for plants. They hold the plant up, and they support the leaves so that they will be in sunlight. Stems also carry water and food to other parts of the plant.

Most plant stems grow upward. The leaves of long-stemmed plants can reach sunlight even in shady places. Some stems even turn during the day. This helps keep the leaves in sunlight, too.

Some plant stems grow sideways, instead of up. Wherever the stem touches the ground, it forms a root from which a new plant grows. Strawberry and spider plants are examples of this type of plant.

Many desert plants have stems that store food and water. The stem of the barrel cactus stores water for the plant. When rain is scarce, the cactus uses water from its stem.

Small plants, such as daisies and dandelions, usually have soft, green, flexible stems. The water inside the stem makes it firm enough to hold the plant up. You might have noticed that a cut flower begins to droop after a few days. Without a root, it can no longer get enough water to keep its stem firm. Most soft-stemmed plants live for just one growing season.

Large plants, such as bushes and trees, need extra support. For this reason, they usually have stiff, woody stems. Woody plants do not die at the end of one growing

◀ This sugar beet root stores food for the plant in the form of sugar. Sugar beets are grown and sold for their sugar.

◀ The sugar cane plant stores sugar in its stems. Sugar cane is also grown and sold for its sugar.

season but continue to grow year after year. Some woody plants, such as the redwoods of California, may live for hundreds or even thousands of years.

Recall that plant stems contain narrow tubes that carry water, nutrients, and food. The tubes that transport water and nutrients are called **xylem** (ZY•luhm). They move water and nutrients upward, from the roots to the leaves. The strings you find when you bite into a stalk of celery are xylem tubes.

The tubes that carry food are called **phloem** (FLOH•em). They move the food made in the leaves to other parts of the plant. The food that plants make and store is needed for growth by roots, stems, and leaves.

In plants with flexible stems, xylem and phloem are in bundles scattered all through the stem. In plants with woody stems, the xylem and phloem are arranged in rings. The xylem is toward the inside of the stem, while the phloem is toward the outside of the stem.

During each growing season, the stem of a woody plant gets thicker as new rings of xylem and phloem form. More xylem than phloem forms each year, so most of the thickness of a tree trunk is xylem. Older rings of xylem no longer transport water. They harden as they become filled with transported materials. This old xylem is the heartwood of a tree. People use the heartwood of many kinds of trees as lumber to build houses and to make furniture.

✓ **Name the two types of tubes that transport materials in plant stems.**

Each year, trees produce a new layer of xylem, forming a *growth ring*. Counting the growth rings of a cut tree can tell you how many years old the tree was.

xylem

phloem

The trunk and branches of a tree are the plant's woody stems. The wood of a tree is old xylem. The bark is old phloem that is pushed outward as the trunk and branches grow thicker. ▶

Leaves

Leaves have many shapes and sizes. Some are smaller than a postage stamp, while others are large enough to cover a school bus. But whether they are big or small, most leaves are thin and flat. This helps them make food.

Leaves are the "food factories" of plants. They use water and nutrients from the soil, carbon dioxide from the air, and energy from sunlight to make food. The food-making process called **photosynthesis** also produces oxygen, which the plants release into the air.

A *pigment*, or coloring matter, called **chlorophyll** (KLAWR•uh•fil) helps plants use light energy to produce sugars. Chlorophyll gives leaves their green color. In the fall, as the days get shorter, most leaves stop making chlorophyll. Then other pigments already in the leaves can be seen. So it's not frost but simply a lack of chlorophyll that makes the beautiful fall colors seen throughout most of the country.

A leaf is not as simple as it may appear. Inside are layers of cells containing microscopic *chloroplasts* (KLAWR•uh•plasts), which are full of chlorophyll. The food-making process takes place inside the chloroplasts. There are also veins, or bundles of xylem and phloem, running through a leaf. Veins bring water and nutrients to the chloroplasts and take sugars from them.

Carbon dioxide enters a leaf, and oxygen and water leave it, through tiny holes called *stomata* (stoh•MAH•tuh). Stomata open wide when the plant has plenty of water. They close to conserve water when necessary. A waxy outer layer on the top of most leaves also helps conserve water.

The leaves or stems of some plants are adapted as *tendrils*. Tendrils wrap themselves around poles or attach themselves to rough surfaces to help the leaves reach the sunlight they need to make food.

Some leaves "catch" food. The Venus' flytrap grows in places where the soil may not have all the nutrients the plant needs to make food. The plant's traplike leaves are adapted to snap shut when an insect lands on them. The leaves release chemicals that digest the insect and take from it the nutrients the plant needs.

Some leaves also store food. The fleshy layers of an onion bulb—the part we eat—are really leaves.

✓ **What is the main thing leaves do?**

The spines (leaves) of a cactus are an adaptation that protects the plant's food and water from desert animals. ▼

Summary

Each part of a vascular plant has a different function. Roots anchor a plant and take in nutrients and water from the soil. A stem supports a plant and moves materials between the plant's parts. Leaves make the plant's food. All of these parts may be adapted to the environment and the needs of the plant.

Review

1. Why are the parts of some plants very different from those of other plants?
2. How are taproots and fibrous roots different?
3. Why do plants store food?
4. **Critical Thinking** What would happen to a green plant if you left it in a dark room for a long time?
5. **Test Prep** Some leaves change color in the fall. This is because —
   A they need to be replaced
   B they stop making chlorophyll
   C there is too much sunlight
   D the trees are dying

LINKS

MATH LINK

Use Variables A ratio is a comparison that uses numbers. If a leaf is 6 cm long and 2 cm wide, the ratio of its length to its width is 3 to 1. Use a ruler to measure the length and width of a number of different leaves. Round each measurement to the nearest centimeter and record your measurements. What observations can you make about the ratio of length to width?

WRITING LINK

Informative Writing—Description Suppose there were a world without trees. Write a story describing what such a world would be like. How would the world be different? What products would be missing from people's lives? Share your story with your classmates.

SOCIAL STUDIES LINK

Map Choose a type of plant and find out where in the United States it grows. Use a computer to make a map and identify the places where the plant grows. Then use the computer to make a chart showing the plant type, location, and climate.

TECHNOLOGY LINK

Learn more about plants by visiting the Harcourt Learning Site.

www.harcourtschool.com

LESSON 2

How Do Plants Reproduce?

In this lesson, you can . . .

 INVESTIGATE nonvascular plants.

 LEARN ABOUT plant reproduction.

 LINK to math, writing, language arts, and technology.

INVESTIGATE

Nonvascular Plants

Activity Purpose Scientists classify plants by the way they transport water. You read in Lesson 1 that the stems of many plants have xylem that carries water from the roots to other parts of the plant. Now you will **observe** plants that have similar-looking parts. You will **infer** what these parts do by **comparing** them to the plant parts you observed in Lesson 1.

Materials
- moss
- liverwort
- hand lens

Activity Procedure

1. **Observe** the moss and the liverwort. **Record** what you see.

2. Now **observe** the plants with a hand lens. Can you see different parts? Do any of the parts you see look like the parts of the potted plant you observed in Lesson 1? (Picture A)

3. **Observe** the plants by touching them with your fingers. Are they soft or firm? Are they dry or moist? What else can you tell by feeling them? Describe what they feel like.

◀ Moss often grows in moist, shady forests. Many tiny plants grow close together to form a mat on tree trunks, rocks, or damp soil.

A98

Picture A

Picture B

4 Touch the plants with a pencil or other object while you **observe** them through the hand lens. Do the parts bend, or are they stiff? Do you see anything new if you push a part of the plant to one side? Describe what you see.

5 **Observe** the plants by smelling them. Do they have any kind of odor? Try to identify the odors. Describe what you smell. (Picture B)

6 Make drawings of the moss and liverwort, identify the parts you observed, and **infer** what each part does.

Draw Conclusions

1. What plant parts did you **observe** on the moss? What parts did you observe on the liverwort?

2. What do you **infer** each part of the plant does?

3. **Scientists at Work** Scientists use observations to **compare** things. Use the observations you made in this investigation to compare the moss and liverwort with a vascular plant.

Investigate Further **Observe** a fern. Based on your observations, would you **classify** a fern as a nonvascular plant, like the moss and the liverwort, or as a vascular plant, like the potted plant in Lesson 1?

Process Skill Tip

By knowing what observations help you **compare** things, you will be able to make better observations.

LEARN ABOUT

FIND OUT

- how nonvascular and vascular plants reproduce

VOCABULARY

spore
gymnosperm
pollen
angiosperm
germinate

The spore capsules of moss plants contain hundreds of tiny spores. Each spore can grow into a new plant. ▼

Different Methods of Reproduction

Nonvascular Plants

Recall from page A52 that mosses and liverworts are simple plants that usually grow in damp places. They need to stay moist because they do not have xylem tubes to transport water. They also lack phloem tubes.

Remember, plants that don't have xylem and phloem are nonvascular plants. Nonvascular plants can move water, nutrients, and food only from one cell to the next. This is the reason why nonvascular plants are so small. Vascular plants, which have xylem and phloem, can grow much larger.

As you observed in the investigation, nonvascular plants have parts that look similar to those of vascular plants. Their leaflike parts, for example, have chloroplasts and use sunlight to manufacture food. Their thin, rootlike structures anchor the plants in the ground and take in some water and nutrients. Their stemlike parts hold the leaflike parts up to the sunlight. However, these similar-looking parts are not true leaves, roots, or stems,

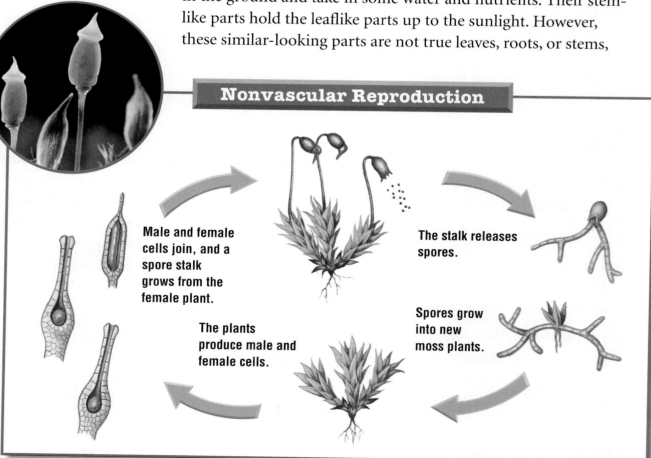

Nonvascular Reproduction

Male and female cells join, and a spore stalk grows from the female plant.

The plants produce male and female cells.

Spores grow into new moss plants.

The stalk releases spores.

because they do not have xylem and phloem.

Nonvascular plants do not have flowers, so they don't reproduce with seeds. Instead, they reproduce with spores (SPOHRZ). A **spore** is a single reproductive cell that grows into a new plant. During their life cycle, mosses produce male and female reproductive cells on separate plants. A male cell and a female cell unite and produce a stalk that grows out of the female plant. The stalk releases the spores that will grow into new moss plants.

✓ **What are the two major groups of plants, and how are they different?**

Simple Vascular Plants

Simple vascular plants include ferns and horsetails. Many people think of ferns as plants with lacy leaves. In fact, there are more than 11,000 kinds of ferns, with many different kinds of leaves.

About 325 million years ago, vast forests of tall tree ferns covered much of the Earth. Today most ferns are found in the tropics, though some grow in cool forests. A few kinds even grow in the Arctic.

Horsetails are much less common than ferns. There are only about 20 kinds of horsetails. Most are small, and all contain silica, a gritty material like sand. Years ago, people used dried horsetails to scrub pots and pans.

Like mosses and liverworts, simple vascular plants reproduce with spores. Also like nonvascular plants, ferns and horsetails have two different stages in their life cycles.

As is the case with mosses, ferns produce male and female reproductive cells. However, in ferns the united cell, or *zygote*, divides and grows into a separate spore-producing plant.

✓ **How do simple vascular plants reproduce?**

◀ The underside of a fern leaf contains spore cases.

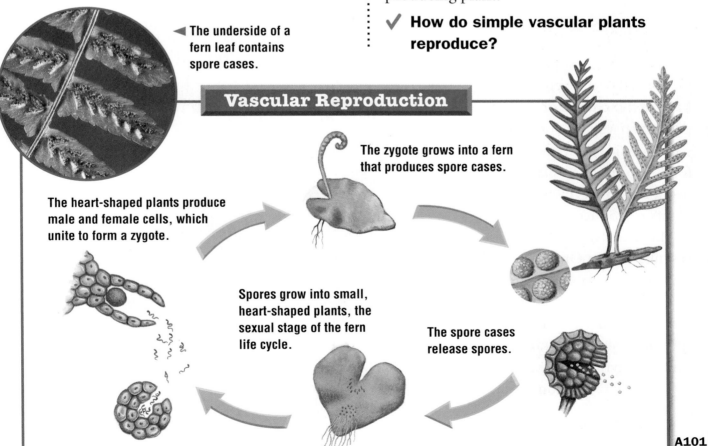

Vascular Reproduction

The heart-shaped plants produce male and female cells, which unite to form a zygote.

The zygote grows into a fern that produces spore cases.

The spore cases release spores.

Spores grow into small, heart-shaped plants, the sexual stage of the fern life cycle.

A101

▲ The cones of spruce trees hang down. The cones of pines grow up.

▶ Conifers, such as this Norway spruce, are common in cold northern climates.

Cone-Bearing Vascular Plants

Spore-producing plants make large numbers of spores. This adaptation makes sure that at least some of the spores will grow into new plants. Seed-producing plants make relatively fewer seeds, but a seed has a better chance of growing into a new plant than a spore does. This is because a seed contains a supply of food. This stored food helps the new plant grow until it can begin making its own food. Most vascular plants reproduce with seeds.

There are two kinds of seed-producing vascular plants. One type produces seeds with no protection. The other type produces seeds protected by some kind of fruit.

Plants with unprotected seeds are called **gymnosperms** (JIM•noh•spermz). The most common gymnosperms are the *conifers* (KAHN•uh•ferz), or cone-bearing plants, such as pine trees.

Most conifers produce both male and female cones on the same tree. Male cones produce **pollen**, structures that contain the male reproductive cells.

Female cones vary in size from 2 cm (about $\frac{3}{4}$ in.) to more than 75 cm (about 2 ft). Their shapes vary, too, but most have a kind of stem from which thin, woody plates grow. These plates are called *scales*.

Wind carries pollen from male cones to female cones. There the male and female reproductive cells unite. The resulting zygotes divide and grow into seeds. During dry weather the scales open and the seeds are released.

✓ **What is a gymnosperm?**

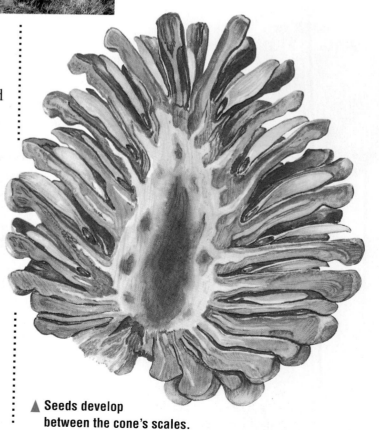

▲ Seeds develop between the cone's scales.

Flowering Vascular Plants

Most of the plants you are familiar with are flowering plants, or **angiosperms** (AN•jee•oh•spermz). There are more than 235,000 kinds of angiosperms on Earth. These include grasses, herbs, shrubs, and many trees. Flowering plants are important sources of wood, fiber, and medicine. Nearly all the food that people eat comes directly or indirectly from flowering plants.

Flowers are an adaptation that is important to the success of angiosperms. They help make sure that pollen gets from the male part of a flower to the female part. Unlike gymnosperms, which are pollinated only by the wind, angiosperms are also pollinated by insects and other small animals. The colors, shapes, and odors of flowers attract these animals, which carry pollen from one flower to another as they move about.

▲ An apple begins as a flower.

◀ There are more than 235,000 kinds of flowering plants. Flowers are important to the success of angiosperms.

Angiosperm seeds are also an adaptation for success. Unlike the gymnosperms, which produce unprotected seeds, angiosperms produce fruits that protect their seeds. These fruits include apples, oranges, tomatoes, peanuts, and acorns.

A fruit protects the seed or seeds inside it in several ways. It usually keeps birds and other animals from getting at them, even if they eat the outer part of the fruit. A fruit also serves as a covering that protects the seeds from cold weather. In addition, a rotting fruit provides extra food for a new plant when it begins to grow.

▲ The seeds of the apple tree are protected by the fruit. As the fruit rots, it provides extra food for the growth of a new apple tree.

✓ **What is an angiosperm?**

Seed Dispersal

Once the eggs of a plant have been fertilized and fruits have formed, the plant is ready to release the seeds. If the fruits fall next to the parent plant, the seeds do not have a good chance of growing. But plants are adapted in many ways to disperse, or scatter, fruits and seeds to places far away from the parent plant.

Maple trees, for example, produce wing-shaped fruits that spin as they fall. Spinning slows down their fall and makes it possible for wind to carry the fruit and its attached seed away from the parent tree.

In the Amazon rain forest, the fruits of some trees are dispersed by dropping into the Amazon River. The fruits are carried down the river, where they may wash up on a distant shore. There the seeds may sprout.

Many plants depend on animals to scatter their seeds. Oak trees, for example, produce fruits called acorns. Squirrels eat some acorns when they fall, and bury others to eat during the winter. The seeds inside buried acorns may sprout and grow into new oak trees.

Some seeds are covered by a fruit called a bur. The outside of a bur is usually rough, and it sticks to the fur of any passing animal. When it finally falls off, it may land on the ground and the seed inside may sprout.

✓ **Name two ways in which plants disperse seeds.**

The rough outside of a bur sticks to the fur of a passing animal or to a human's clothes.

Each dandelion seed is attached to a bit of fluff that can be blown by a light wind.

The seeds of this berry will be left in some other place with the bird's droppings.

When an animal eats a fruit, the seed may fall to the ground and sprout.

Seed Germination

A seed survives inside its protective seed coat until conditions are right for it to grow. These conditions usually include fertile soil, warm temperatures, and enough rainfall or moisture. Most seeds can survive for several years, and some seeds have survived for hundreds of years. When conditions are right, a seed will sprout, or **germinate** (JER•mih•nate).

First, the seed takes in water. This makes the seed larger. As the seed swells, the seed coat splits. The *embryo*, or tiny plant within the seed, then begins to grow and develop the parts it needs to live on its own. The first part to develop is the root, which begins to grow down, toward the center of the Earth.

Next, the stem emerges from the seed and begins to grow up, toward the light. The seed leaves are attached to the stem. At this stage, the growing plant, now called a *seedling,* uses food stored in the seed leaves to grow. Later the first true leaves, which have also emerged from the seed, will begin to make food.

As it grows, the seedling produces longer and thicker roots. The stem gets taller and stronger. When the seedling is growing well and its leaves are making all the food the plant needs, the seed leaves drop off. The young, rapidly growing plant can now live on its own.

✓ **What is the first part to emerge from a germinating seed?**

▲ This seed has landed on rain-soaked, fertile soil. It will take in moisture, swell, and germinate.

▲ The first part to emerge from the seed is the root.

▲ As the root gets longer and thicker, a stem begins to emerge.

▲ The seedling now has a well-developed root system, and its first true leaves are producing food.

Comparing Life Cycles

Both animals and plants go through stages in their lives. A flowering plant sprouts from a seed, grows and matures, flowers, and produces seeds of its own. An animal is born, grows into an adult, and reproduces its own kind. Each organism completes a cycle of life.

Some young animals look very much like their parents. Puppies and kittens are very small when they are born, but you can easily see what they will grow up to become.

Other young animals look very different from their parents. Who would guess that a caterpillar becomes a beautiful butterfly or that a fishlike tadpole becomes a frog?

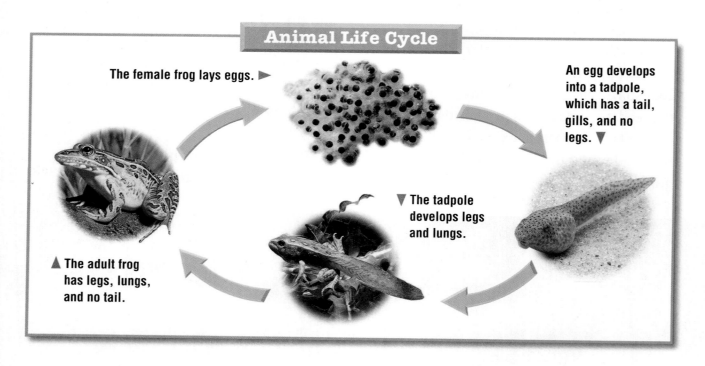

Animal Life Cycle

- The female frog lays eggs.
- An egg develops into a tadpole, which has a tail, gills, and no legs.
- The tadpole develops legs and lungs.
- The adult frog has legs, lungs, and no tail.

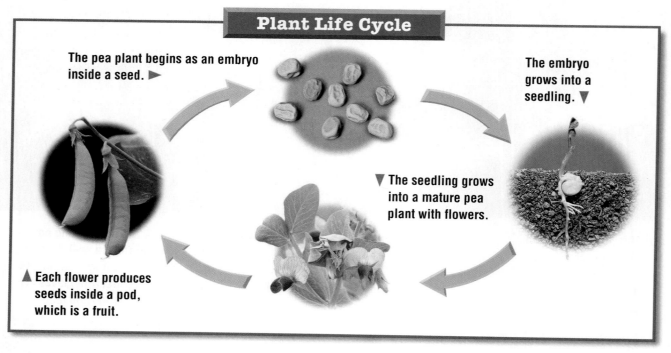

Plant Life Cycle

- The pea plant begins as an embryo inside a seed.
- The embryo grows into a seedling.
- The seedling grows into a mature pea plant with flowers.
- Each flower produces seeds inside a pod, which is a fruit.

In its earliest stages, a flowering plant is also very different from its parent plants. It starts as an embryo inside a seed. When the seed sprouts, the embryo grows into a seedling that does not look much like the mature plant. As the plant grows and matures, it looks more and more like the plants it came from.

✓ **How does the life cycle of a flowering plant compare to that of an animal?**

Summary

Vascular plants have xylem and phloem. Nonvascular plants do not have these tubes. Nonvascular plants and simple vascular plants reproduce with spores. Gymnosperms and angiosperms are seed-producing vascular plants. Like animals, plants go through several stages in their life cycles.

Review

1. Why are nonvascular plants so small?
2. How do ferns reproduce?
3. How do conifers produce seeds?
4. **Critical Thinking** Why do you think night-blooming plants have less colorful flowers than day-blooming plants?
5. **Test Prep** The fruit produced by an angiosperm —
   A makes the seeds taste better
   B protects the seeds inside
   C attracts birds and insects
   D is more attractive than the cone of a gymnosperm

MATH LINK

Display Data Take a look at the plants around you—at home, at school, in parks. Are most of them vascular or nonvascular? Are they gymnosperms or angiosperms? Use a computer graphing program to make a circle graph that compares the percentages of the types of plants you find.

WRITING LINK

Informative Writing—Explanation Write a paragraph explaining how you use plants each day. How many times do you use them? For what purposes? Would it be hard to get through a day without plants? Share your paragraph with your classmates.

LANGUAGE ARTS LINK

Prefixes In the word *nonvascular,* the prefix *non-* means "not." What do you think the words *nonsense, nonbreakable,* and *nonfat* mean? What other *non-* words can you think of? Make a list of words that begin with *non-* and write down what each one means.

TECHNOLOGY LINK

Learn more about flowering plants by visiting the Smithsonian Institution Internet site.
www.si.edu/harcourt/science

Smithsonian Institution®

LESSON 3

How Do People Use Plants?

In this lesson, you can . . .

 INVESTIGATE how heat and moisture can change popcorn.

 LEARN ABOUT the many uses of plants.

 LINK to math, writing, music, and technology.

How many uses of plants or plant products do you see here? ▽

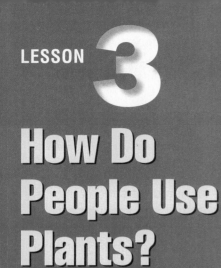

Popcorn

Activity Purpose People eat many kinds of seeds as food, but popcorn is probably the most interesting, and the most fun. Popped popcorn is the exploded seeds of a type of corn plant. Popcorn seeds contain water, although you can't see it or feel it. Heating the seeds turns the water quickly to steam. As the steam expands, the popcorn seeds pop. In this investigation you will **predict** and **measure** how popping the seeds affects their volume and mass.

Materials
- large plastic measuring cup
- unpopped popcorn
- balance

Activity Procedure

1. Cover the bottom of the measuring cup with unpopped popcorn seeds.

2. **Estimate** the volume of the unpopped seeds. Put the cup on the balance, and **measure** the mass of the unpopped seeds. (Picture A)

3. **Predict** what will happen to the mass and the volume when the seeds are popped.

Picture A

Picture B

4 Your teacher will help you pop the popcorn. Return the popped seeds to the measuring cup.

5 **Measure** the volume and mass of the cup of popped popcorn. Were your **predictions** correct? (Picture B)

Draw Conclusions

1. How did the volume of the popcorn change?
2. How did the mass change? Explain.
3. **Scientists at Work** One reason why scientists **experiment** is to test predictions. If an experiment doesn't turn out the way they predicted, it may mean that their predictions were wrong. Or it may mean that they did not consider everything that could affect the experiment. Did you predict the volume and mass of the popped popcorn correctly? Explain.

Investigate Further What other questions do you have about popcorn? **Plan and conduct an experiment** to answer your questions.

Process Skill Tip

If you aren't careful when you **experiment**, something you may not have considered may affect your results.

LEARN ABOUT

The Uses of Plants

Plants as Food

FIND OUT

- how people use plants as food
- how people use plants as medicine

VOCABULARY

grain
fiber

People use plants more for food than for any other purpose. For example, breakfast cereal is made of **grain**, or the seeds of certain grasses. If you have a sandwich for lunch, you are eating grain again. The bread in the sandwich was made by grinding the seeds of wheat into flour. Does your sandwich have lettuce and tomato on it? Then you're also eating a plant leaf and a fruit. And if the sandwich is seasoned with mustard, you're eating something made from seeds.

People eat many different parts of many different plants. Beans, lentils, corn, and rice, for example, are seeds. Beets, radishes, turnips, and carrots are roots. Bamboo shoots and asparagus are stems. Spinach, lettuce, kale, and cabbage are leaves. Cherries, pears, oranges, and olives are fruits. Artichokes, cauliflower, and broccoli are flowers. And if you like cinnamon in your apple pie, you are eating the bark, the outer part of the stem, of a tree.

✓ **Name the plant parts that people eat.**

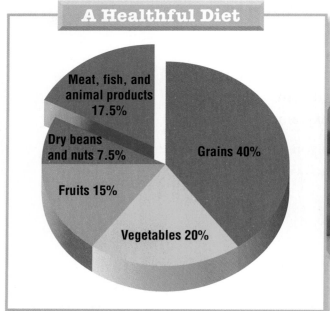

In a healthful diet, most of the foods come from plants.

THE INSIDE STORY

The Food Guide Pyramid

Grains form the largest part of this pyramid because they are the foundation of a healthful diet. The next level is shared by vegetables and fruits. Drinking fruit juice is more healthful than drinking soda, but you should also eat the fruits themselves.

Meats and dairy products are in a small part of the pyramid because they contain fats. Too much fat can harm your health. Fish and poultry have less fat than other meats. Milk, cheese, and yogurt are good for you, but they also are high in fat. Foods that are very high in fat, oil, or sugar are not healthful.

FATS, OILS, SWEETS
Foods in this group are generally not healthful, so don't eat a lot of them.

MILK, YOGURT, CHEESE GROUP
These foods contain fats, so limit yourself to 2–3 servings a day. A cup of milk is one serving.

MEAT, POULTRY, FISH, DRY BEANS, EGGS, NUTS GROUP
Eat 2–3 servings of foods from this group daily. A serving is one egg or about 3 oz of meat.

VEGETABLE GROUP
Eat 3–5 servings of vegetables a day. A half cup of chopped vegetables is one serving.

FRUIT GROUP
Eat 2–4 servings a day. A banana is one serving.

BREAD, CEREAL, RICE, PASTA GROUP
Eat 6–11 servings a day. A slice of bread, for example, is one serving.

◀ The leaves of the aloe plant store food. That food, a jellylike substance, is used in soaps, shampoos, makeup, skin creams, and sunscreens.

This fluffy cotton boll is made of tiny white fibers. The fibers are woven into cotton cloth, which is made into clothing. ▼

Plants as Medicines

Plants contain many substances that can be used to treat illnesses. Native Americans used the leaves and roots of hundreds of plants as medicines. They used them to reduce fevers, relieve pain, calm upset stomachs, and treat other problems.

About 40 percent of the medicines we use today are made from plants. For example, an important heart medicine called digitalis is made from the leaves of the foxglove plant. Foxglove grows in many parts of the United States. Quinine is made from the bark of a tree that grows in the Andes Mountains of South America. Quinine is used to treat malaria.

One of the best-known and most widely used pain medicines is also one of the oldest. Aspirin is a medicine invented in the 1800s. But thousands of years earlier, people took an almost identical medicine by chewing the bark of the willow tree.

✓ **What did some people do for pain before aspirin was invented?**

Other Uses for Plants

Clothing is another important product people get from plants. Blue jeans, for example, are made of fibers from the cotton plant. A **fiber** is any material that can be separated into thread. The dye that gives blue jeans their color was once made from the indigo plant.

Many kinds of trees provide wood for different purposes. Homes are often made of wood, and a lot of the furniture in most homes is wood. Musical instruments, such as guitars, violins, and pianos, are made with wood. And pulp, which is made from wood, is used to make paper.

Soaps and shampoos contain plant substances that can help make skin smooth and hair shiny. Many perfumes are made from flower petals. It takes about 100 kg (220 lb) of rose petals to make 30 mL (about 1 oz) of fragrance. This is one of the reasons why perfumes are expensive.

✓ **Name two products made from trees.**

This cancer medicine was once made from the bark of the Pacific Yew. ▼

Summary

People eat the leaves, stems, roots, seeds, fruits, and flowers of various plants. When they are sick, people often use medicines made from plants. In fact, many things people use every day come from plants.

Review

1. Name three foods that are seeds or are made from seeds.
2. From which food group should you eat the most servings each day?
3. What percentage of the medicines people use comes from plants?
4. **Critical Thinking** Nutritionists say people shouldn't eat many French fries. In which food groups do French fries belong? What part of this food may not be healthful?
5. **Test Prep** Aloe is a plant that is used to make —
   A dye
   B skin cream
   C aspirin
   D fiber

LINKS

MATH LINK

Collect Data Most people should eat no more than 30 g of fat each day. Food labels list the fat in each serving. Add up the fat in the food you eat in one week. How close to 30 g per day is the amount of fat in your diet?

WRITING LINK

Persuasive Writing—Business Letter Some people think it's healthful to eat only plants. Others think it's important to eat both meat and plants. Write a letter to a health organization requesting information about the reasons for and against both of these diets.

MUSIC LINK

Wooden Instruments Today, some musical instruments are still made of wood. Report to the class on one of them. Explain why it is made of wood instead of some other material. Include a picture of the instrument, and play part of a recording of the instrument.

TECHNOLOGY LINK

Learn more about using plants for food by viewing *Genetic Tomatoes* on the **Harcourt Science Newsroom Video** in your classroom video library.

SCIENCE AND TECHNOLOGY

Corn Cards and Super Slurpers

Chemists and agricultural scientists are working together to find new ways to use plants. Their goal is to make useful products that won't cause pollution.

Why Use Plants to Make Plastic?

Products made from plastics make life simpler, but plastics can cause problems, too. Most plastics are made from petroleum, and when you throw them away, they aren't really gone. Petroleum-based plastics don't decompose (break down) in the environment. Each year people throw away almost 20 million tons of plastics. That's a lot of trash. Now scientists who work with plants have discovered how to use corn to produce plastics that do decompose. In a landfill or a

Corn is one of America's most commonly grown crops.

backyard compost pile, these plastics break down into hydrogen, oxygen, and small bits of organic matter called humus. Humus helps enrich the soil.

Diapers? Not As Corny As It Sounds

What are corn-based plastics good for? Plenty! Companies use these environmentally friendly materials to make products ranging from disposable diapers to prepaid telephone calling cards.

Packaging "peanuts" made from cornstarch protect fragile products. After they have been exposed to moisture and sunlight for a few months, they fall apart.

Chemists at a U.S. Department of Agriculture laboratory discovered a way to use cornstarch to make a material they called hydrosorb. Its nickname is "Super Slurper." Why? Because it can soak up more than 300 times its own weight in water. Diapers containing hydrosorb help keep babies dry. Super Slurper filters remove water from fuels such as gasoline and heating oil. When mixed with soil, hydrosorb holds moisture near the roots of plants, helping them grow with less irrigation.

PLA Plastics

Other scientists have found ways to recombine the hydrogen, oxygen, and carbon in corn to make a material called polylactic acid resin, or PLA. Plastics made from PLA can be used just like petroleum-based plastics to make toys, TV sets, and other products.

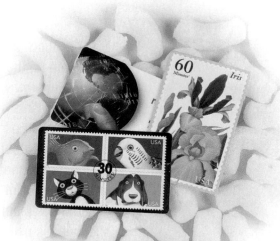
Packaging "peanuts" and phone cards can be made from corn.

Nebraska farmers grow nearly 30 million tons of corn each year. Scientists at the University of Nebraska say that they haven't found anything made with petroleum-based plastic that can't also be made from plastic that comes from corn.

THINK ABOUT IT

1. What other products could Super Slurpers be used for?
2. What are some advantages and disadvantages of corn-based plastics?

CAREERS
AGRONOMIST

What They Do Agronomists study soils and plants to develop better ways to grow crops and to keep agricultural land productive.

Education and Training People wishing to become agronomists study plant biology and soil chemistry in college and graduate school.

 WEB LINK
For Science and Technology updates, visit the Harcourt Internet site.
www.harcourtschool.com

PEOPLE IN SCIENCE

"Our food, many of the things we wear, and even some parts of our homes come from plants. People also get enjoyment from looking at plants."

Shirley Mah Kooyman
BOTANIST

Because we depend on plants for so many things, the health of plants is important to our survival. Shirley Mah Kooyman feels that plants are sometimes taken for granted—that people forget they are living things. Ms. Kooyman is a botanist, or plant scientist, in Chanhassen, Minnesota. An important part of her work is discovering what makes plants grow. The knowledge that she gains is used by other scientists, who work to find ways to grow healthier plants that produce larger crops.

In her work, Ms. Kooyman seeks to better understand some of the growth processes of plants. She knows that in addition to light energy, water, proper temperature, and rich soil, plants need certain hormones. Hormones are chemical "messengers" that "tell" plants to grow. Plant hormones are produced in stems and roots. From there they travel to other parts of the plant.

Today scientists are able to make artificial plant hormones. Artificial hormones placed on a root tip make the root grow, just as natural hormones do. The advantage of artificial hormones is that they can be produced in greater amounts than natural hormones. Farmers can use artificial hormones to speed up plant growth and produce larger crops. This will make plant products more affordable. In addition to working with plant hormones in a laboratory and in the field, Ms. Kooyman teaches people about plants and the joys of gardening.

THINK ABOUT IT

1. Why is it important to understand what makes plants grow?
2. How might artificial hormones be used to produce a large crop of tomatoes?

ACTIVITIES FOR HOME OR SCHOOL

WATER IN PLANTS

How does water move through plants?

Materials
- 5 toothpicks
- dropper
- water

Procedure

1. Break the toothpicks in half, but don't separate the parts. The two halves should remain connected.

2. Arrange the toothpicks like the spokes in a wagon wheel.

3. Put several drops of water in the center of the "wheel."

4. Observe any changes to the toothpicks.

Draw Conclusions

What happened to the water you put on the toothpicks? What happened to the toothpicks? Relate this to the way water moves through plants.

LEAF CASTS

How can you observe stomata?

Materials
- potted plant
- clear fingernail polish
- microscope slide
- microscope

Procedure

1. Paint a 2-cm square of fingernail polish on the underside of one leaf. Let the polish dry.

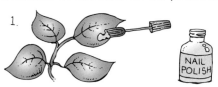
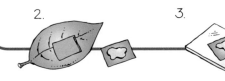

2. Add another layer of polish and let it dry. Repeat until you have 6 layers of polish.

3. Once the last layer of polish is dry, peel it off the leaf.

4. Put the polish, which contains a cast of the leaf epidermis, on a microscope slide.

5. Observe the slide by using the microscope.

Draw Conclusions

You should observe two types of cells. Compare the cells. Explain any differences between them. How do the guard cells form the stomata?

CHAPTER 4 Review and Test Preparation

Vocabulary Review

Use the terms below to complete the paragraph. The page numbers in ( ) tell you where to look in the chapter if you need help.

xylem (A95)
phloem (A95)
photosynthesis (A96)
chlorophyll (A96)
spores (A101)
gymnosperms (A102)
pollen (A102)
angiosperms (A103)
germinate (A105)

A plant that contains tubes for the transport of water and food is called a vascular plant. The tubes that transport food are **1.** tubes; those that transport water are **2.** tubes. Plants that do not have these tubes are nonvascular plants. Whether or not they have transport tubes, all plants have a green pigment called **3.**, which enables them to make their own food. As plants make food, they use carbon dioxide and release oxygen. This process is called **4.**.

Plants have various means by which they reproduce. Plants will produce either seeds or **5.**, which will grow into a new plant when conditions are favorable. Conifers, or **6.**, produce unprotected seeds in cones. The flowers of **7.** produce protected seeds. In a flower, **8.** contains the male reproductive cells. When conditions are right, a seed will sprout, or **9.**.

Connect Concepts

Write terms from the Word Bank to complete the chart.

angiosperms   seeds
gymnosperms   spores
nonvascular plants   vascular plants

The Two Main Groups of Plants Are

10. _____, which have no xylem or phloem.
11. _____, which have xylem and phloem tubes.

The Two Groups of Vascular Plants Are

12. plants that use _____ to reproduce.
13. plants that use _____ to reproduce.

The Two Groups of Seed-Producing Plants Are

14. _____, which include conifers, such as pines.
15. _____, which include flowering apple trees.

Check Understanding

Write the letter of the best choice.

16. Most leaves are thin and flat because —
    A they look better that way
    B this helps them make food
    C they protect the plant from insects and birds
    D they absorb water from the air

17. Nonvascular plants are limited in size because —
    F they must pass water and nutrients from one cell to the next
    G they do not make their own food
    H birds and other animals like to eat them
    J they live in shady places

18. Large, colorful flowers are useful for a plant because —
    A they look pretty
    B they make food for plants
    C they attract insects and birds that spread pollen
    D they collect moisture

19. For a more healthful diet, eat more of foods that are —
    F near the top of the Food Guide Pyramid
    G mostly from the milk group
    H high in sugar
    J near the bottom of the Food Guide Pyramid

Critical Thinking

20. Could a plant live if all its leaves were cut off? Explain your answer.

21. In what way are fruits better than cones for carrying seeds?

22. How does successful seed dispersal help to ensure that a plant species will survive?

Process Skills Review

23. What can you **observe** about a plant to help you identify the kind of plant it is?

24. **Infer** which products in your classroom are made from plants or plant parts.

25. **Predict** what would happen if trees did not have fruits or cones.

Performance Assessment

Design a Plant

Choose one condition from each set. Design a plant with roots, stems, and leaves that could live in a place with the conditions you choose.

a lot of rain
some rain
almost no rain

hot temperatures
moderate temperatures
cold temperatures

a lot of light
some light
almost no light

UNIT A EXPEDITIONS

There are many places where you can observe living systems. By visiting the places below, you can learn about some of the differences among living things. You'll also have fun while you learn.

Huntsville-Madison County Botanical Garden

WHAT An ornamental garden displaying a wide assortment of plant species
WHERE Huntsville, Alabama
WHAT CAN YOU DO THERE? Take the Garden Tour, explore the trails, and study the adaptations of the many plants.

Butterfly Pavilion & Insect Center

WHAT An insect zoo that is home to more than a thousand spectacular butterflies
WHERE Westminster, Colorado
WHAT CAN YOU DO THERE? Observe subtropical plants and see different kinds of butterflies from around the world.

GO ONLINE Plan Your Own Expeditions

If you can't visit the Huntsville-Madison County Botanical Garden or the Butterfly Pavilion & Insect Center, visit a garden or zoo near you. Or log on to The Learning Site at **www.harcourtschool.com** to visit these science sites and other places where you can observe living systems.

References

Science Handbook

Using Science Tools — **R2**

    Using a Hand Lens — R2
    Using a Thermometer — R2
    Caring for and Using a Microscope — R3
    Using a Balance — R4
    Using a Spring Scale — R4
    Measuring Liquids — R5
    Using a Ruler or Meterstick — R5
    Using a Timing Device — R5
    Measurement Systems — R6

Health Handbook — **R7**

Using Science Reading Strategies — **R38**

Building a Science Vocabulary — **R44**

Glossary — **R50**

Index — **R60**

Using Science Tools

Using a Hand Lens

A hand lens magnifies objects, or makes them look larger than they are.

1. Hold the hand lens about 12 centimeters (5 in.) from your eye.
2. Bring the object toward you until it comes into focus.

Using a Thermometer

A thermometer measures the temperature of air and most liquids.

1. Place the thermometer in the liquid. Don't touch the thermometer any more than you need to. Never stir the liquid with the thermometer. If you are measuring the temperature of the air, make sure that the thermometer is not in line with a direct light source.
2. Move so that your eyes are even with the liquid in the thermometer.
3. If you are measuring a material that is not being heated or cooled, wait about two minutes for the reading to become stable. Find the scale line that meets the top of the liquid in the thermometer, and read the temperature.
4. If the material you are measuring is being heated or cooled, you will not be able to wait before taking your measurements. Measure as quickly as you can.

Caring for and Using a Microscope

A microscope is another tool that magnifies objects. A microscope can increase the detail you see by increasing the number of times an object is magnified.

Caring for a Microscope

- Always use two hands when you carry a microscope.
- Never touch any of the lenses of a microscope with your fingers.

Using a Microscope

1. Raise the eyepiece as far as you can using the coarse-adjustment knob. Place your slide on the stage.
2. Always start by using the lowest power. The lowest-power lens is usually the shortest. Start with the lens in the lowest position it can go without touching the slide.
3. Look through the eyepiece, and begin adjusting it upward with the coarse-adjustment knob. When the slide is close to being in focus, use the fine-adjustment knob.
4. When you want to use a higher-power lens, first focus the slide under low power. Then, watching carefully to make sure that the lens will not hit the slide, turn the higher-power lens into place. Use only the fine-adjustment knob when looking through the higher-power lens.

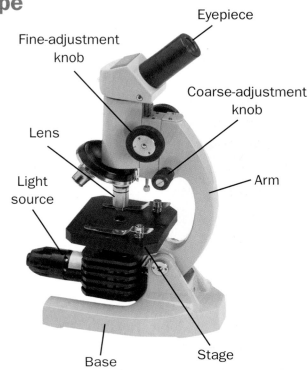

A Light Microscope

You may use a Brock microscope. This is a sturdy microscope that has only one lens.

1. Place the object to be viewed on the stage.
2. Look through the eyepiece, and begin raising the tube until the object comes into focus.

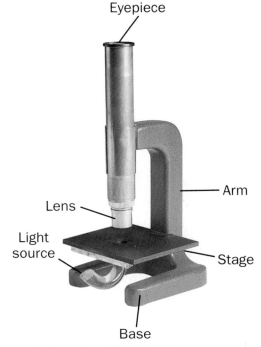

A Brock Microscope

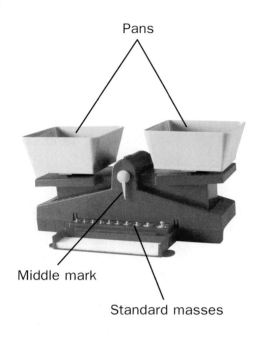

Pans

Middle mark

Standard masses

Using a Balance

Use a balance to measure an object's mass. Mass is the amount of matter an object has.

1. Look at the pointer on the base to make sure the empty pans are balanced.
2. Place the object you wish to measure in the left pan.
3. Add the standard masses to the other pan. As you add masses, you should see the pointer move. When the pointer is at the middle mark, the pans are balanced.
4. Add the numbers on the masses you used. The total is the mass in grams of the object you measured.

Using a Spring Scale

Use a spring scale to measure forces such as the pull of gravity on objects. You measure weight and other forces in units called newtons (N).

Measuring the Weight of an Object

1. Hook the spring scale to the object.
2. Lift the scale and object with a smooth motion. Do not jerk them upward.
3. Wait until any motion of the spring comes to a stop. Then read the number of newtons from the scale.

Measuring the Force to Move an Object

1. With the object resting on a table, hook the spring scale to it.
2. Pull the object smoothly across the table. Do not jerk the object.
3. As you pull, read the number of newtons you are using to pull the object.

Measuring Liquids

Use a beaker, a measuring cup, or a graduate to measure liquids accurately.

1. Pour the liquid you want to measure into a measuring container. Put your measuring container on a flat surface, with the measuring scale facing you.
2. Look at the liquid through the container. Move so that your eyes are even with the surface of the liquid in the container.
3. To read the volume of the liquid, find the scale line that is even with the surface of the liquid.
4. If the surface of the liquid is not exactly even with a line, estimate the volume of the liquid. Decide which line the liquid is closer to, and use that number.

Beaker **Graduate**

Using a Ruler or Meterstick

Use a ruler or meterstick to measure distances and to find lengths of objects.

1. Place the zero mark or end of the ruler or meterstick next to one end of the distance or object you want to measure.
2. On the ruler or meterstick, find the place next to the other end of the distance or object.
3. Look at the scale on the ruler or meterstick. This will show the distance you want or the length of the object.

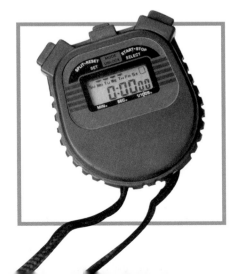

Using a Timing Device

Use a timing device such as a stopwatch to measure time.

1. Reset the stopwatch to zero.
2. When you are ready to begin timing, press start.
3. As soon as you are ready to stop timing, press stop.
4. The numbers on the dial or display show how many minutes, seconds, and parts of seconds have passed.

Measurement Systems

SI Measures (Metric)

Temperature
Ice melts at 0 degrees Celsius (°C)
Water freezes at 0°C
Water boils at 100°C

Length and Distance
1000 meters (m) = 1 kilometer (km)
100 centimeters (cm) = 1 m
10 millimeters (mm) = 1 cm

Force
1 newton (N) = 1 kilogram ×
    1 meter/second/second (kg-m/s^2)

Volume
1 cubic meter (m^3) = 1m × 1m × 1m
1 cubic centimeter (cm^3) =
    1 cm × 1 cm × 1 cm
1 liter (L) = 1000 milliliters (mL)
1 cm^3 = 1 mL

Area
1 square kilometer (km^2) =
    1 km × 1 km
1 hectare = 10,000 m^2

Mass
1000 grams (g) = 1 kilogram (kg)
1000 milligrams (mg) = 1 g

Rates (Metric and Customary)
km/h = kilometers per hour
m/s = meters per second
mi/h = miles per hour

Customary Measures

Volume of Fluids
8 fluid ounces (fl oz) = 1 cup (c)
2 c = 1 pint (pt)
2 pt = 1 quart (qt)
4 qt = 1 gallon (gal)

Temperature
Ice melts at 32 degrees
    Fahrenheit (°F)
Water freezes at 32°F
Water boils at 212°F

Length and Distance
12 inches (in.) = 1 foot (ft)
3 ft = 1 yard (yd)
5280 ft = 1 mile (mi)

Weight
16 ounces (oz) = 1 pound (lb)
2000 pounds = 1 ton (T)

Health Handbook

Nutrition and Food Safety
The Food Guide Pyramid	R8
Understanding Serving Size	R9
Fight Bacteria	R10
Food Safety Tips	R11

Being Physically Active
Planning Your Weekly Activities	R12
Guidelines for a Good Workout	R13
Warm-Up and Cool-Down Stretches	R14
Building a Strong Heart and Lungs	R16

Safety and First Aid
Fire Safety	R18
Earthquake and Thunderstorm Safety	R19
First Aid for Choking and Bleeding	R20

Caring for Your Body Systems
Sense Organs	R22
Skeletal System	R24
Muscular System	R26
Digestive System	R28
Circulatory System	R30
Respiratory System	R32
Nervous System	R34
Immune System	R36
Endocrine System	R37

Good Nutrition

The Food Guide Pyramid

No one food or food group supplies all the nutrients you need. That's why it's important to eat a variety of foods from all the food groups. The Food Guide Pyramid can help you choose healthful foods in the right amounts. By choosing more foods from the groups at the bottom of the pyramid, and few foods from the group at the top, you will eat nutrient-rich foods that provide your body with energy to grow and develop.

Fats, oils, and sweets Eat sparingly.

Meat, poultry, fish, dried beans, eggs, and nuts 2–3 servings

Milk, yogurt, and cheese 2–3 servings

Fruits 2–4 servings

Vegetables 3–5 servings

Breads, cereals, rice, and pasta 6–11 servings

Understanding Serving Size

The Food Guide Pyramid suggests a number of servings to eat each day from each group. But a serving isn't necessarily the amount you eat at a meal. A plate full of macaroni and cheese may contain three or four servings of pasta (macaroni) and three servings of cheese. That's about half your bread group servings and all your milk servings at one sitting! The table below can help you estimate the number of servings you are eating.

Food Group	Amount of Food in One Serving	Easy Ways to Estimate Serving Size
Bread, Cereal, Rice, and Pasta Group	$\frac{1}{2}$ cup cooked pasta, rice, or cereal 1 ounce ready-to-eat (dry) cereal 1 slice bread, $\frac{1}{2}$ bagel	ice-cream scoop large handful of plain cereal or a small handful of cereal with raisins and nuts
Vegetable Group	1 cup of raw, leafy vegetables $\frac{1}{2}$ cup other vegetables, cooked or chopped raw $\frac{3}{4}$ cup vegetable juice $\frac{1}{2}$ cup tomato sauce	about the size of a fist ice-cream scoop
Fruit Group	medium apple, pear, or orange $\frac{1}{2}$ large banana, or one medium banana $\frac{1}{2}$ cup chopped or cooked fruit $\frac{3}{4}$ cup of fruit juice	about the size of a baseball
Milk, Yogurt, and Cheese Group	$1\frac{1}{2}$ ounces of natural cheese 2 ounces of processed cheese 1 cup of milk or yogurt	about the size of two dominoes $1\frac{1}{2}$ slices of packaged cheese
Meat, Poultry, Fish, Dried Beans, Eggs, and Nuts Group	3 ounces of lean meat, chicken, or fish 2 tablespoons peanut butter $\frac{1}{2}$ cup of cooked dried beans	about the size of your palm
Fats, Oils, and Sweets Group	1 teaspoon of margarine or butter	about the size of the tip of your thumb

Preparing Foods Safely

Fight Bacteria

You probably already know to throw away food that smells bad or looks moldy. But food doesn't have to look or smell bad to make you ill. To keep your food safe and yourself from becoming ill, follow the procedures shown in the picture below. And remember—when in doubt, throw it out!

Food Safety Tips

Tips for Preparing Food

- Wash hands in hot, soapy water before preparing food. It's also a good idea to wash hands after preparing each dish.
- Defrost meat in the microwave or the refrigerator. Do NOT defrost meat on the kitchen counter.
- Keep raw meat, poultry, fish, and their juices away from other food.
- Wash cutting boards, knives, and countertops immediately after cutting up meat, poultry, or fish. Never use the same cutting board for meats and vegetables without thoroughly washing the board first.

Tips for Cooking Food

- Cook all food thoroughly, especially meat. Cooking food completely kills bacteria that can make you ill.
- Red meats should be cooked to a temperature of 160°F. Poultry should be cooked to 180°F. When done, fish flakes easily with a fork.
- Eggs should be cooked until the yolks are firm. Never eat food that contains raw eggs. Never eat cookie dough made with raw eggs.

Tips for Cleaning Up the Kitchen

- Wash all dishes, utensils, and countertops with hot, soapy water. Use a disinfectant soap, if possible.
- Store leftovers in small containers that will cool quickly in the refrigerator. Don't leave leftovers on the counter to cool.

Being Physically Active

Planning Your Weekly Activities

Being active every day is important for your overall health. Physical activity strengthens your body systems and helps you manage stress and maintain a healthful weight. The Activity Pyramid, like the Food Guide Pyramid, can help you choose a variety of activities in the right amounts to keep your body strong and healthy.

The Activity Pyramid

Sitting for more than thirty minutes at a time: Only Once in a While

Flexibility and Strength: Two to Three Times a Week

Light Exercise: Two to Three Times a Week

Twenty-plus minutes of continuous aerobic activity: Three to Five Times a Week

Stay active: Every Day

Guidelines for a Good Workout

There are three things you should do every time you are going to exercise—warm up, workout, and cool down.

Warm-Up When you warm up, your heart rate, respiration rate, and body temperature gradually increase and more blood begins to flow to your muscles. As your body warms up, your flexibility increases, helping you avoid muscle stiffness after exercising. People who warm up are also less prone to exercise-related injuries. Your warm-up should include five minutes of stretching, and five minutes of a low-level form of your workout exercise. For example, if you are going to run for your primary exercise, you should spend five minutes stretching, concentrating on your legs and lower back, and five minutes walking before you start running.

Workout The main part of your exercise routine should be an aerobic exercise that lasts twenty to thirty minutes. Some common aerobic exercises include walking, bicycling, jogging, swimming, cross-country skiing, jumping rope, dancing, and playing racket sports. You should choose an activity that is fun for you and that you will enjoy doing over a long period of time. You may want to mix up the types of activities you do. This helps you work different muscle groups, and provides a better overall workout.

Cool-Down When you finish your aerobic exercise, you need to give your body time to return to normal. You also need to stretch again. This portion of your workout is called a cool-down. Start your cool-down with three to five minutes of low-level activity. For example, if you have been running, you may want to jog and then walk during this time. Then do stretching exercises to prevent soreness and stiffness.

Being Physically Active

Warm-Up and Cool-Down Stretches

Before you exercise, you should always warm up your muscles. The warm-up stretches shown here should be held for at least fifteen to twenty seconds and repeated at least three times. At the end of your workout, spend about two minutes repeating some of these stretches.

▶ **Hurdler's Stretch** HINT—Keep the toes of your extended leg pointed up.

▲ **Shoulder and Chest Stretch** HINT—Pulling your hands slowly toward the floor makes this stretch more effective. Keep your elbows straight, but not locked!

▼ **Sit-and-Reach Stretch** HINT—Remember to bend at the waist. Keep your eyes on your toes!

HEALTH HANDBOOK

▲ **Upper Back and Shoulder Stretch**
HINT—Try to stretch your hand down so that it lies flat against your back.

▼ **Thigh Stretch** HINT—Keep both hands flat on the floor. Try to lean as far forward as you can.

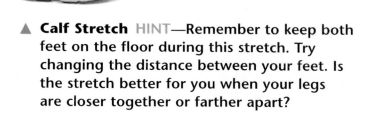

▲ **Calf Stretch** HINT—Remember to keep both feet on the floor during this stretch. Try changing the distance between your feet. Is the stretch better for you when your legs are closer together or farther apart?

Tips for Stretching

- Never bounce when stretching.
- Remember to hold each stretch for fifteen to twenty seconds.
- Breathe normally. This helps your body get the oxygen it needs.
- Stretch only until you feel a slight pull, NOT until it hurts.

Being Physically Active

Building a Strong Heart and Lungs

Aerobic activities, those that cause deep breathing and a fast heart rate for at least twenty minutes, help both your heart and your lungs. Because your heart is a muscle, it gets stronger with exercise. A strong heart doesn't have to work as hard to pump blood to the rest of your body. Exercise also allows your lungs to hold more air. With a strong heart and lungs, your cells get oxygen faster and your body works more efficiently.

◀ **Swimming** Swimming may provide the best overall body workout of any sport. It uses all the major muscle groups, and improves flexibility. The risk of injury is low, because the water supports your weight, greatly reducing stress on the joints. Just be sure to swim only when a lifeguard is present.

▶ **In-Line Skating** In-line skating gives your heart and lungs a great workout. Remember to always wear a helmet when skating. Always wear protective pads on your elbows and knees, and guards on your wrists, too. Learning how to skate, stop, and fall correctly will make you less prone to injuries.

HEALTH HANDBOOK

▶ **Tennis** To get the best aerobic workout from tennis, you should run as fast, far, and hard as you can during the game. Move away from the ball so that you can step into it as you hit it. Finally, try to involve your entire body in every move.

▲ **Bicycling** Bicycling provides good aerobic activity that places little stress on the joints. It's also a great way to see the countryside. Be sure to use a bike that fits and to learn and follow the rules of the road. And *always* wear your helmet!

▶ **Walking** A fast-paced walk is a terrific way to build your endurance. The only equipment you need is a good pair of shoes and clothes appropriate for the weather. Walking with a friend can make this exercise a lot of fun.

Safety in Emergencies

Fire Safety

Fires cause more deaths than any other type of disaster. But a fire doesn't have to be deadly if you prepare your home and follow some basic safety rules.

- Install smoke detectors outside sleeping areas and on any additional floors of your home. Be sure to test the smoke detectors once a month and change the batteries in each detector twice a year.

- Keep a fire extinguisher on each floor of your home. Check monthly to make sure each is properly charged.

- Work with your family to make a fire escape plan for each room of your home. Ideally, there should be two routes out of each room. Sleeping areas are most important, because most fires happen at night. Plan to use stairs only; elevators can be dangerous in a fire.

- Pick a place outside for everyone to meet. Designate one person to call the fire department or 911 from a neighbor's home.

- Practice crawling low to avoid smoke. If your clothes catch fire, follow the three steps listed below.

1. STOP
2. DROP
3. ROLL

Earthquake Safety

An earthquake is a strong shaking of the ground. The tips below can help you and your family stay safe in an earthquake.

Before an Earthquake	During an Earthquake	After an Earthquake
• Secure tall, heavy furniture, such as bookcases, to the wall. Store the heaviest items on the lowest shelves.	• If you are outdoors, stay outdoors and move away from buildings and utility wires.	• Continue to watch for falling objects as aftershocks shake the area.
• Check for potential fire risks. Bolt down gas appliances, and use flexible hosing and connections for both gas and water utilities.	• If you are indoors, take cover under a heavy desk or table or in a doorway. Stay away from glass doors and windows and heavy objects that might fall.	• Adults should have the building checked for hidden structural problems.
• Reinforce and anchor overhead light fixtures to help keep them from falling.	• If you are in a car, drive to an open area away from buildings and overpasses.	• Check for broken gas, electric, and water lines. If you smell gas, an adult should shut off the gas main and leave the area. Report the leak.

Thunderstorm Safety

Thunderstorms are severe storms. Lightning associated with thunderstorms can injure or kill people, cause fires, and damage property. Here are some thunderstorm safety tips.

- **If you are inside, stay there.** The best place to take cover is inside a building.
- **If you are outside, try to take shelter.** If possible, get into a closed car or truck. If you can't take shelter, crouch in a ditch or low area, if possible.
- **If you are outside, stay away from tall objects.** Don't stand under a lone tree, in an open field, on a beach, or on a hilltop. Find a low place to stay.
- **Stay away from water.** Lightning is attracted to water, and water conducts electricity.
- **Listen for weather bulletins and updates.** The storms that produce lightning may also produce tornadoes. Be ready to take shelter in a basement or interior hallway away from windows and doors.

First Aid

The tips on the next few pages can help you provide simple first aid to others or yourself. Always tell an adult about any injuries that occur.

For Choking . . .

If someone else is choking . . .

  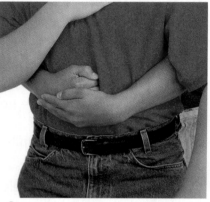

1. Recognize the Universal Choking Sign—grasping the throat with both hands. This sign means a person is choking and needs help.

2. Stand behind the choking person, and put your arms around his or her waist. Place your fist above the person's belly button. Grab your fist with your other hand.

3. Pull your hands toward yourself, and give five quick, hard, upward thrusts on the person's stomach.

If you are choking when alone . . .

1. Make a fist, and place it above your belly button. Grab your fist with your other hand. Pull your hands up with a quick, hard thrust.

2. Or, keep your hands on your belly, lean your body over the back of a chair or over a counter, and shove your fist in and up.

For Bleeding . . .

If someone else is bleeding . . .

Wash your hands with soap, if possible.

Put on protective gloves, if available.

Wash small wounds with soap and water. Do *not* wash serious wounds.

Place a clean gauze pad or cloth over the wound. Press firmly for ten minutes. Don't lift the gauze during this time.

If you don't have gloves, have the injured person hold the gauze or cloth in place with his or her hand for ten minutes.

If after ten minutes the bleeding has stopped, bandage the wound. If the bleeding has not stopped, continue pressing on the wound and get help.

If you are bleeding . . .

- Wash your wound if it is a small cut. If it is a serious wound, do *not* wash it.
- Place a gauze pad or clean cloth over the wound, and hold it firmly in place for ten minutes. Don't lift the gauze or cloth until ten minutes have passed.
- If you have no gauze or cloth, apply pressure with your hand.
- If after ten minutes the bleeding has stopped, bandage the wound. If the bleeding has not stopped, continue pressing on the wound and get help.

HEALTH HANDBOOK

Sense Organs

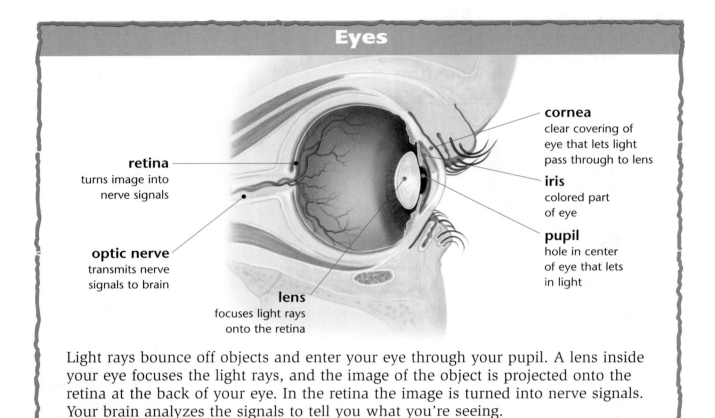

Light rays bounce off objects and enter your eye through your pupil. A lens inside your eye focuses the light rays, and the image of the object is projected onto the retina at the back of your eye. In the retina the image is turned into nerve signals. Your brain analyzes the signals to tell you what you're seeing.

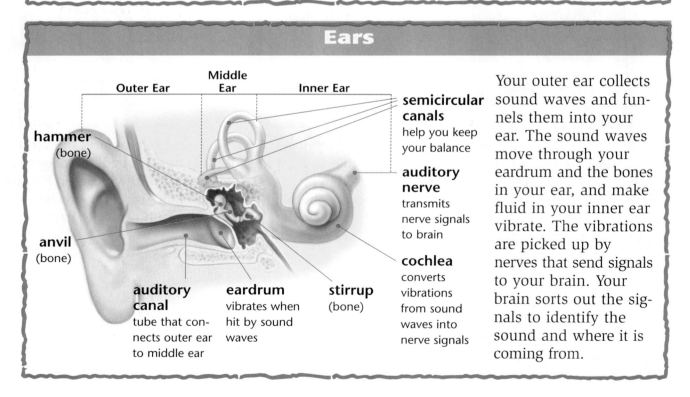

Your outer ear collects sound waves and funnels them into your ear. The sound waves move through your eardrum and the bones in your ear, and make fluid in your inner ear vibrate. The vibrations are picked up by nerves that send signals to your brain. Your brain sorts out the signals to identify the sound and where it is coming from.

Nose

- **skull**
- **nasal bone**
- **olfactory (sense of smell) bulb** — group of nerves that carry information to olfactory tract
- **nostrils** — openings to nose
- **olfactory tract** — carries information from olfactory bulb to brain
- **mucous membrane** — warms and moistens air you breathe in
- **nasal cavity** — main opening inside nose

When you breathe in, air is swept upward to nerve cells in your nasal cavity. Your nasal cavity is the upper part of your nose inside your skull. Different nerve cells respond to different odors in the air and send signals to your brain.

Skin

The skin is made of three layers, the outer epidermis, the middle dermis, and the lower subcutaneous layer. Nerve cells in your skin signal your brain about stimuli (conditions around you) that affect your skin.

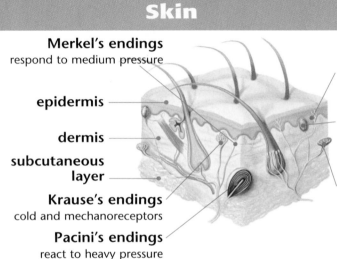

- **Merkel's endings** — respond to medium pressure
- **epidermis**
- **dermis**
- **subcutaneous layer**
- **Krause's endings** — cold and mechanoreceptors
- **Pacini's endings** — react to heavy pressure
- **free nerve endings** — react to painful stimuli
- **Meissner's endings** — respond to light pressure and small, fast vibration
- **Ruffini's endings** — sense changes in temperature and pressure

Caring for Your Senses

Injuries to your brain can affect your senses. Protect your brain by wearing safety belts in the car and helmets when playing sports or riding your bike.

Tongue

Your tongue is covered with about 10,000 tiny nerve cells, or taste buds, that pick out tastes in the things you eat and drink. Different taste buds respond to different tastes and send signals to your brain.

- **taste buds**

Activity

With a partner, toss a table tennis ball back and forth 15 times. Then put a patch over one eye and toss the ball again. Was it easier to catch the ball when you had both eyes open or only one eye open?

Skeletal System

Your bones fit together and attach to your muscles at your joints. Your bones and muscles work together at your joints to allow you to move in many directions. Each joint is designed to do a job.

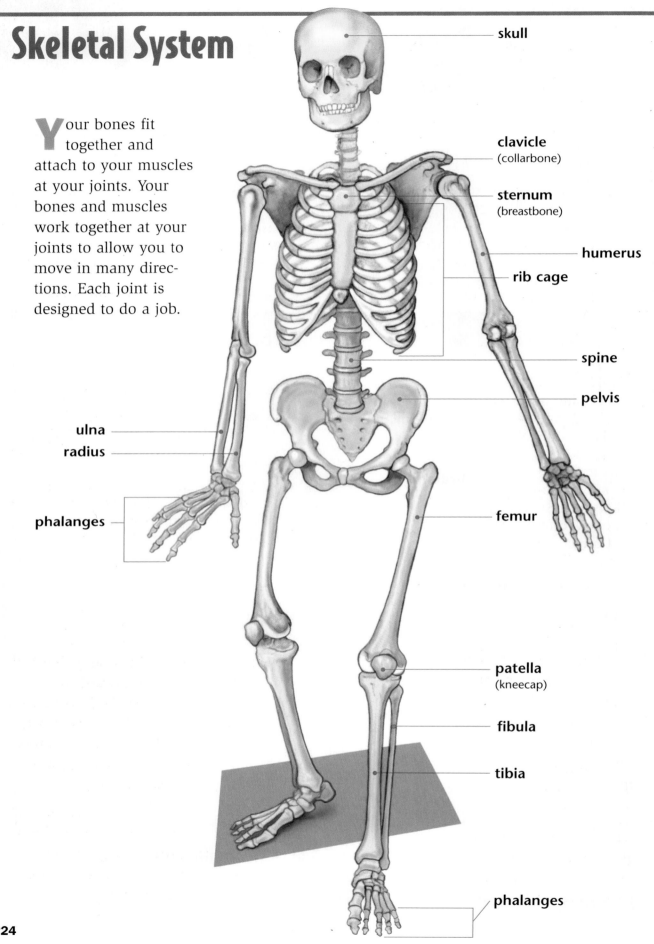

HEALTH HANDBOOK

Bones and Joints

Kinds of Bones Bones come in four basic shapes: long, short, flat, and irregular. Long bones, such as those in your legs, arms, and fingers, are narrow with large ends and are slightly curved. These bones can support the most weight. Short bones, found in your wrists and ankles, are chunky and wide. These bones allow maximum movement around a joint. Flat bones, such as your skull and ribs, are platelike. They provide protection for especially delicate parts of your body. Irregular bones, such as those in your spine and your ears, have unique shapes that don't fit into any other category.

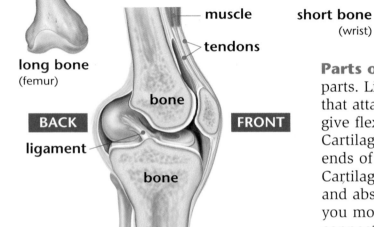

flat bone (rib)

irregular bone (vertebra)

short bone (wrist)

long bone (femur)

Parts of a Joint A joint has several parts. Ligaments are tough, elastic bands that attach one bone to another. Ligaments give flexibility for bending and stretching. Cartilage is a cushioning material at the ends of bones that meet in a joint. Cartilage helps the bones move smoothly and absorbs some of the impact when you move. Tendons are dense cords that connect bones to muscles.

Caring for Your Skeletal System

- Move and flex your joints regularly through exercise. If you don't, they might get injured or become stiff and sore. Be sure to warm up and cool down whenever you exercise.
- You can injure a joint by using it too much or moving it in a way it is not designed to move.
- Calcium is necessary for healthy bones. You can get calcium from milk, dairy products like yogurt and cheese, or some dark green, leafy vegetables.

Activities

1. Build a model of a long bone and a short bone. Use a sheet of construction paper. Cut off a strip about 1 inch (2.5 cm) wide. Roll the strip and the remainder of the sheet into tight cylinders. Test their strength by putting objects on top of them.

2. Make a model of a joint. Cut out two strips of cardboard and join them together using a round metal fastener. What does the fastener represent?

Muscular System

The muscles that make your body move are attached to bones. When one of these muscles contracts, or gets shorter, it pulls on the bone it's connected to and the bone moves.

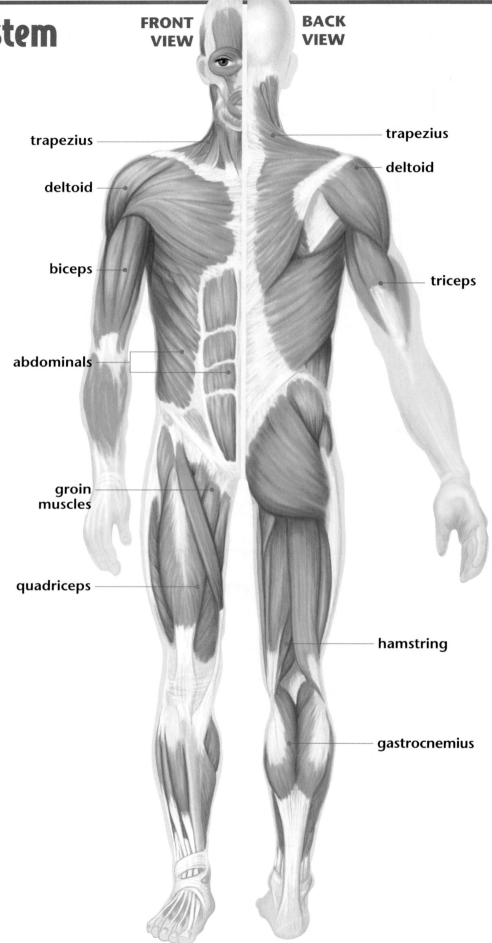

HEALTH HANDBOOK

Muscles and Bones

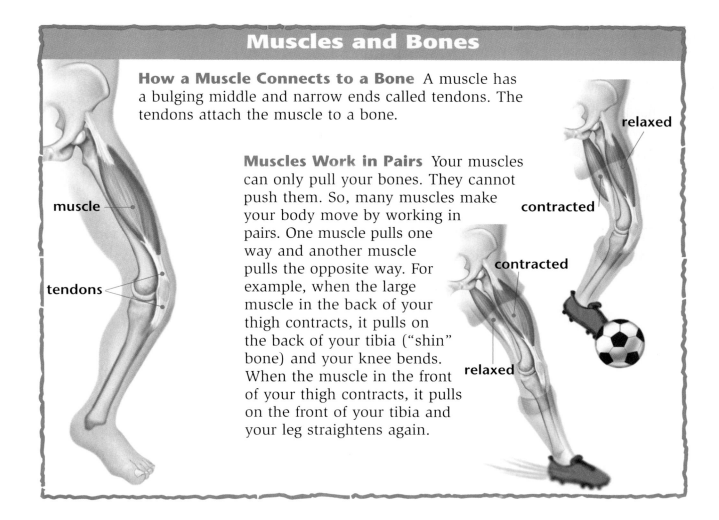

How a Muscle Connects to a Bone A muscle has a bulging middle and narrow ends called tendons. The tendons attach the muscle to a bone.

Muscles Work in Pairs Your muscles can only pull your bones. They cannot push them. So, many muscles make your body move by working in pairs. One muscle pulls one way and another muscle pulls the opposite way. For example, when the large muscle in the back of your thigh contracts, it pulls on the back of your tibia ("shin" bone) and your knee bends. When the muscle in the front of your thigh contracts, it pulls on the front of your tibia and your leg straightens again.

Caring for Your Muscular System

- Take a brisk five-minute walk and do gentle stretches before you start to play a sport to loosen your tendons, ligaments, and muscles and help prevent injuries.
- You should ease out of exercise too. Because your muscles contract during exercise, they need to be stretched when you finish.

Activities

1. Wrap your fingers around your upper arm, so you can feel both the top and bottom of your arm. Slowly bend and straighten your arm several times. Which muscles are working as a pair?
2. Write a paragraph about which muscles work in pairs when you play your favorite sport.

Digestive System

Digestion is the process of breaking food into tiny pieces that are absorbed by your blood and carried to all parts of your body. Each part of your digestive system does a different job.

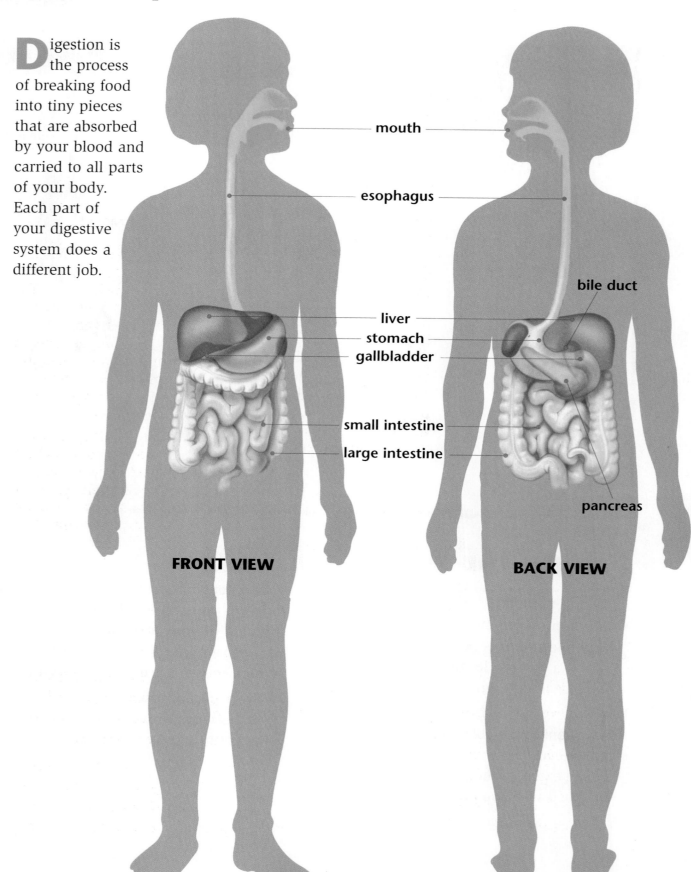

- mouth
- esophagus
- liver
- stomach
- gallbladder
- small intestine
- large intestine
- bile duct
- pancreas

FRONT VIEW   **BACK VIEW**

HEALTH HANDBOOK

Some Digestive Organs

Esophagus When you swallow, the chewed and moistened food goes down a tube to your stomach. This tube, called the esophagus, is about 10 inches (25.4 cm) long. The food is pushed down the esophagus by a squeezing muscle action, similar to squeezing a tube of toothpaste. When you swallow, your airway is protected by the epiglottis and your vocal cords. If you try to talk and swallow, you cough or choke.

Stomach The walls of your stomach are made of very strong muscles. Once food reaches the stomach, more chemicals are added that digest the food more thoroughly. At the same time, the stomach muscles squeeze the liquid food mixture.

Food leaves your stomach in two stages. First, the top of your stomach squeezes the liquid food mixture into the small intestine. Second, the solid particles are pushed into the small intestine by the muscles in the lower part of your stomach.

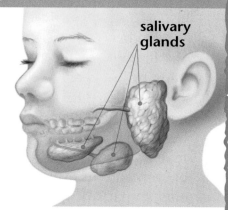

Salivary Glands When you chew your food, it is moistened with saliva. The saliva comes from three sets of glands: one set under your tongue, one on each side of your head in front of your ear, and one on each side of your head under your jaw.

Saliva is mostly water, but it does contain a chemical called *amylase* that helps digest starches. Saliva also helps keep your mouth clean and helps control infection.

Caring for Your Digestive System

- Try to drink at least six to eight glasses of water a day. Water helps food move through your digestive system.
- Chew your food thoroughly. Large pieces of food are more difficult to digest, and the large pieces might cause you to choke.

Activities

1. Chew a cracker and then hold it in your mouth for about a minute. Move it around. How has the taste changed?
2. Record the amount of water you drink every day for three days. Are you drinking at least six glasses of water every day?

Circulatory System

Blood is carried throughout your body by arteries, veins, and capillaries. Arteries deliver blood with needed materials, such as oxygen and nutrients, to parts of your body. Veins carry blood with waste or unused materials, such as carbon dioxide. Capillaries are microscopic blood vessels that allow needed substances to seep into your body's tissues.

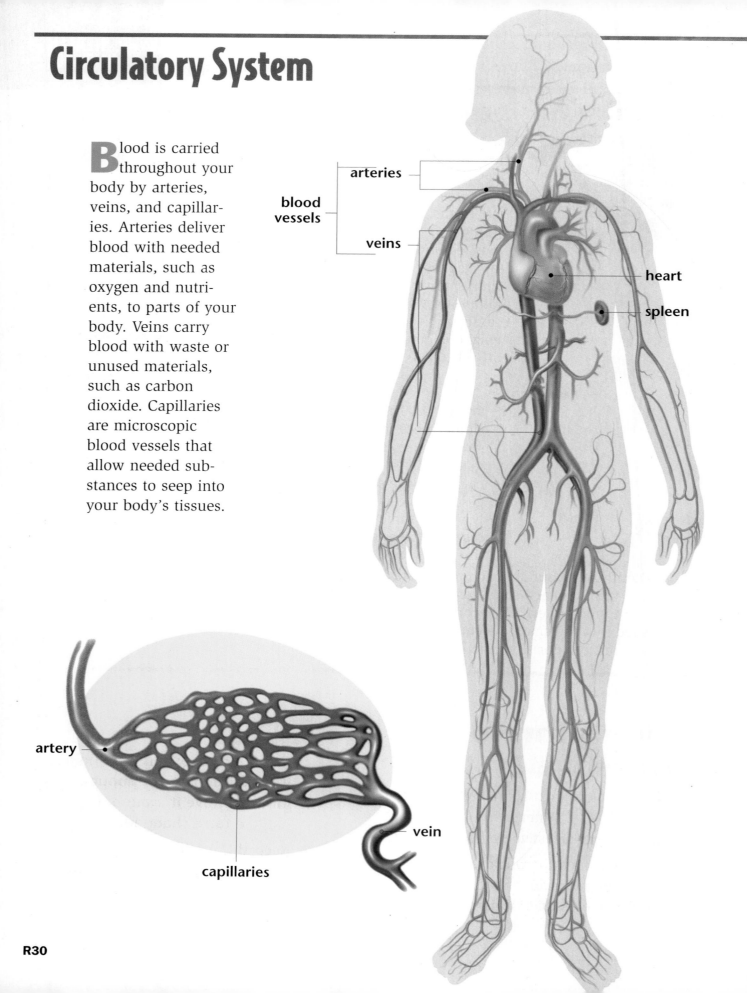

HEALTH HANDBOOK

How Blood Moves in Your Heart

Your heart has four chambers. The upper chambers are called atria. The lower chambers are called ventricles. Here is how blood moves through the heart.

1. Blood comes from your body into the right atrium. Blood comes from your lungs into the left atrium. The atria squeeze.
2. Two valves open, and the blood moves into the ventricles. When both ventricles have filled with blood, the valves shut. The ventricles squeeze.
3. Two different valves then open. The right ventricle sends blood through one valve to the lungs. The left ventricle sends blood through the other valve to the body. The "lub-dub" of your heartbeat is the sound of valves slamming shut.

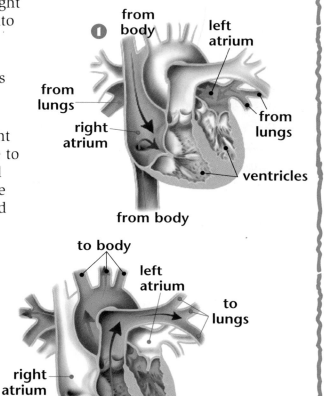
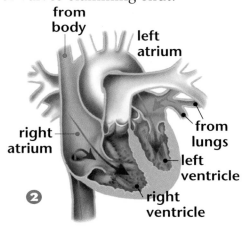

Caring for Your Circulatory System

- Don't ever smoke. Smoking narrows blood vessels and can cause high blood pressure.
- Remember, your heart is a constantly working muscle. Exercise strengthens your heart by making it beat harder, which makes the heart muscles larger and able to push more blood with each "squeeze" or "beat."

Activities

1. Take your pulse for ten seconds. Multiply that number by six to find how many times your heart beats in a minute. How many times does your heart beat in a day?
2. Feel your pulse in various places—your wrist, your neck, behind your knee. Where is it easiest to feel?

Respiratory System

Your lungs are filled with air tubes, air sacs, and blood vessels. The air tubes and blood vessels in your lungs divide until they are very small. At the ends of the tiny air tubes are air sacs called alveoli. The smallest blood vessels, capillaries, surround the alveoli.

Blood coming from your body (blue) delivers waste gases to your lungs. It then picks up oxygen (red) and takes it to your body.

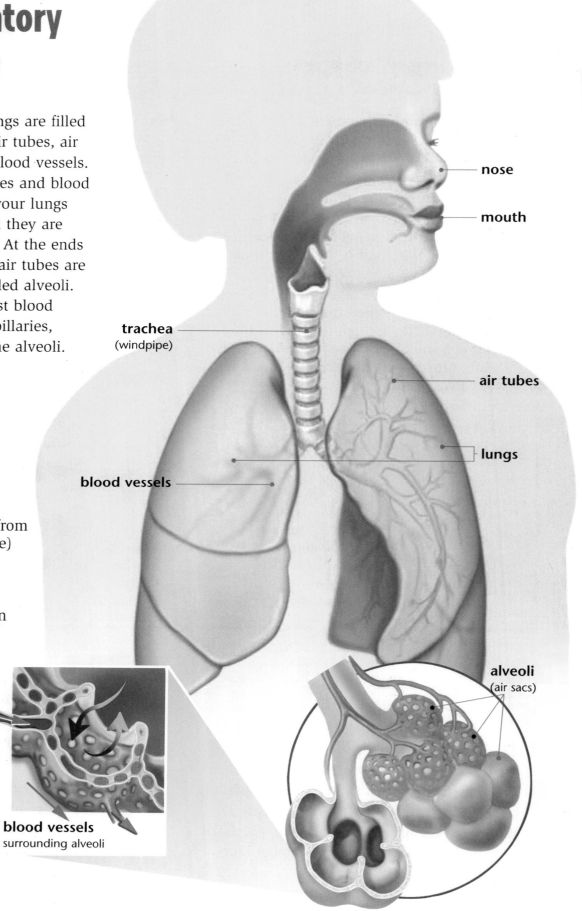

HEALTH HANDBOOK

How Oxygen Travels to Your Body Parts

Your heart and lungs are connected by veins and arteries. The blood your heart pumps out to the rest of your body comes directly from your lungs, where it is filled with oxygen. The blood delivers the oxygen and picks up carbon dioxide. When the blood returns to your heart, it needs oxygen. Your heart sends the blood to your lungs, where oxygen is added and carbon dioxide is removed.

Tiny blood vessels in your lungs release carbon dioxide into the alveoli and absorb oxygen from the air inside the alveoli. The replenished blood returns to the heart, and the process starts over.

It takes just one minute for blood to circulate around your entire body.

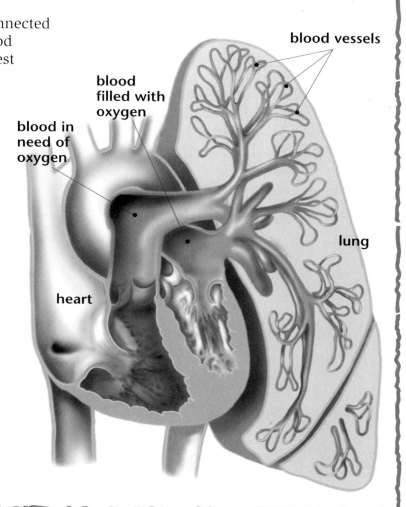

Caring for Your Respiratory System

- Don't ever smoke. The tar in cigarettes damages the lungs. Avoid environmental tobacco smoke too. Inhaling someone else's smoke can be as dangerous as smoking a cigarette yourself, especially for people with asthma.
- If you exercise so hard that you can't talk or you feel dizzy, your body is not getting enough oxygen. Slow down.

Activities

1. List three sports you think would exercise your respiratory system the most and three sports you think would exercise it the least.
2. Survey people you know about their thoughts on cigarettes and health. Record their answers.
3. Write the words to a jingle for an anti-smoking commercial.

Nervous System

Your nervous system is responsible for all of your thoughts and your body's activity. It makes your heart beat and your lungs work. It allows you to see, hear, smell, taste, and touch. It lets you learn, remember, and feel emotions. It moves all of your muscles.

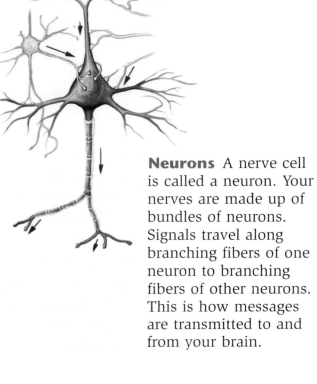

Neurons A nerve cell is called a neuron. Your nerves are made up of bundles of neurons. Signals travel along branching fibers of one neuron to branching fibers of other neurons. This is how messages are transmitted to and from your brain.

Your Nervous System

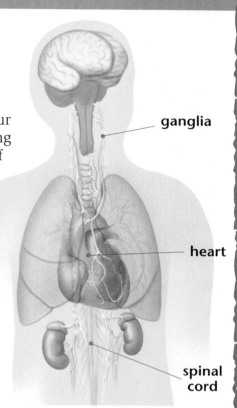

Autonomic Nervous System Your autonomic nervous system keeps your body's systems functioning and in balance. Masses of nerves called ganglia receive messages from your brain and relay those messages to organs like your heart, lungs, and kidneys.

Peripheral Nervous System Major nerves branch off your brain and spinal cord. They continue to branch and form a complicated network that spreads throughout your body. This is your peripheral nervous system.

Caring for Your Nervous System

- Do not take any drugs unless given by your parents or guardian or a doctor. Some drugs can affect your brain cells.
- Eat a well-balanced diet. Your nervous system cannot work properly without certain nutrients.
- Learning new skills builds new nerve connections in your brain.

Activities

1. Put one hand in a bowl of very warm water and one hand in a bowl of cold water. Then put both hands in a bowl of warm water. How does each hand feel?

2. Blindfold a friend. Touch your friend's back with two fingers and your friend's palm with two fingers. Can your friend tell how many fingers you used in each place?

Immune System and Endocrine System

Your immune system defends your body from harmful invaders, such as organisms that cause infection. White blood cells, which are your immune system's primary infection fighters, are produced in your bone marrow, thymus, lymph nodes, and spleen.

Tonsils Your tonsils are also lymph tissue. They produce white blood cells. When your body has an infection, lymph nodes can become infected. Infected lymph nodes become swollen and painful.

tonsils
thymus
lymph nodes
spleen
long bone marrow

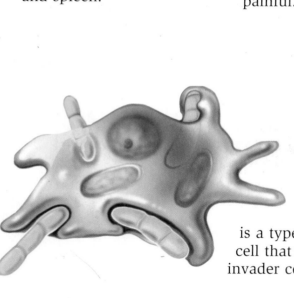

Macrophage A macrophage is a type of white blood cell that eats harmful invader cells.

Caring for Your Immune System

One way to help your body fight disease is through immunization (shots that protect you from certain diseases). You were required to have some immunizations as an infant and "boosters" before you started school.

Activity

Find out what immunizations are recommended at various ages. What will you need next?

HEALTH HANDBOOK

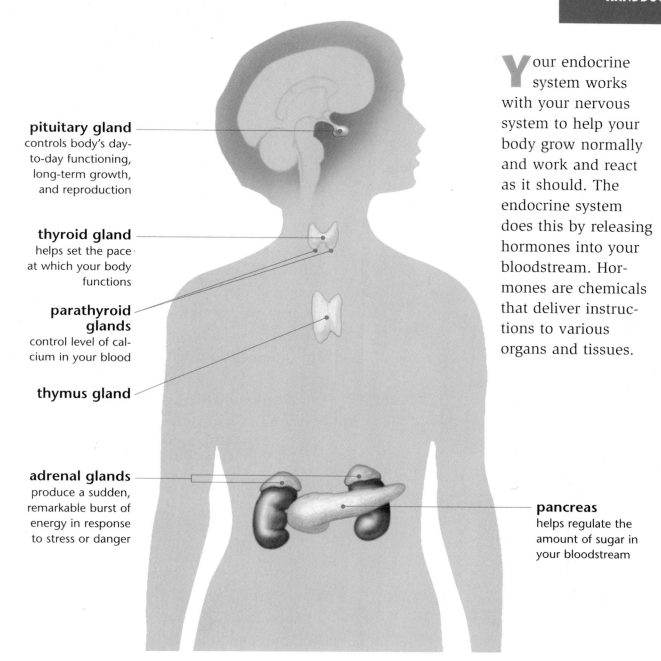

pituitary gland controls body's day-to-day functioning, long-term growth, and reproduction

thyroid gland helps set the pace at which your body functions

parathyroid glands control level of calcium in your blood

thymus gland

adrenal glands produce a sudden, remarkable burst of energy in response to stress or danger

pancreas helps regulate the amount of sugar in your bloodstream

Your endocrine system works with your nervous system to help your body grow normally and work and react as it should. The endocrine system does this by releasing hormones into your bloodstream. Hormones are chemicals that deliver instructions to various organs and tissues.

Caring for Your Endocrine System

Some drugs, such as steroids, can harm your endocrine system. Avoid these drugs.

Activity

The next time you are startled, feel your heartbeat. How is it different from normal?

Using Science Reading Strategies

Some strategies work better than others when it comes to reading science topics. Below are some good strategies to use, and how you might use them.

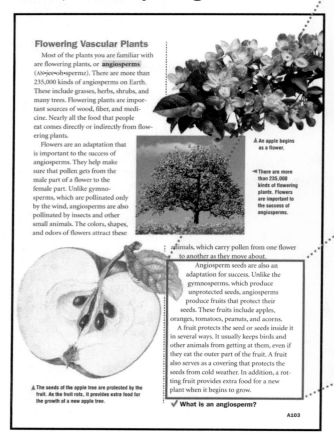

A good strategy to use is...
USE PRIOR KNOWLEDGE

Think about questions like these:
- What new information are you learning?
- How does the information fit in with what you already know about this scientific topic?

When reading a text like this...

> Angiosperm seeds are also an adaptation for success. Unlike the gymnosperms, which produce unprotected seeds, angiosperms produce fruits that protect their seeds. These fruits include apples, oranges, tomatoes, peanuts, and acorns.
>
> A fruit protects the seed or seeds inside it in several ways. It usually keeps birds and other animals from getting at them, even if they eat the outer part of the fruit. A fruit also serves as a covering that protects the seeds from cold weather. In addition, . . .
>
> **page A103**

What you might say to yourself...

"I know that fruit contains seeds. I've seen apple seeds and tomato seeds. I also know that these seeds can grow into apple trees and tomato plants. But I've always thought that fruit was just a kind of food for people and animals. I never thought about what purpose fruit had for the plant itself. Now that I think of it, though, growing fruit around its seeds is the perfect way for a plant to make sure that new seeds are safe and have a chance to grow."

READING STRATEGIES

When reading a text like this...

> ### Activity Purpose
> In Lesson 1 you learned about three forces that affect objects on Earth: gravitation, friction, and magnetism. Two or more of these forces often interact, or act at the same time. In this investigation you will experiment with two opposing forces—the force of gravity pulling down on an object and the force of a spring pulling up on the same object.
>
> ### Materials
> - clipboard
> - graph paper
> - tape
> - ring stand
> - spring
> - weight
> - marker
>
> ### Activity Procedure
>
> **1** Tape the graph paper to the clipboard. Across the bottom of the graph paper, draw a line and label it *Seconds*. Starting at one end of the line, make a mark every 2.5 cm.
>
> **2** Attach the spring to the ring stand. Then attach the weight to the free end of the spring. Tape the marker to the bottom of the weight so that its tip points toward the back of the setup.
>
> page F10

A good strategy to use is...

ADJUST READING RATE

Think about the kind of science text you are reading.

- Something that is new to you or complicated should be read slowly.
- Something that is familiar can be read more quickly.

What you might say to yourself...

"I see three kinds of text on this page: The first paragraph describes the reason for doing the activity. The next kind of text is a list of what I'll need for the investigation. The last kind of text tells me what I need to do, step by step.

"I can probably read the description once and read it pretty quickly—just to get an idea of the reason for doing the activity.

"I'll glance through the list quickly, even though I may look back at it more than once. I want to make sure I have everything I need.

"I'll have to slow down and carefully read all the directions. Then I'll have to reread and check them a few times. I want to be sure I do the investigation the right way."

When reading a text like this...	A good strategy to use is...	What you might say to yourself...
The Causes of Weather **Uneven Heating** The illustration below shows how the sun's rays strike the Earth's surface and the atmosphere. The atmosphere absorbs some of the sun's energy and reflects some of it back into space. Some of the energy that reaches the Earth's surface is reflected back into the atmosphere. However, much of the sun's energy is absorbed by the Earth's surface. *page C72*	**SELF-QUESTION** Think about questions you could ask yourself to be sure you understand what you're reading. As you read your science textbook, one way to check yourself is to turn headings into questions that you answer as you read.	"I can turn both of these headings into questions that I'll try to answer as I read. My first question would be, 'What are the causes of weather?' I'll wait to fully answer that question when I finish the whole section. "The question I'll answer first, after I read a few paragraphs, comes from the second heading. That question is, 'How does uneven heating cause weather?'"
Think about an in-line skater holding a basketball. If the skater throws the ball really hard, the ball goes forward and the skater rolls backward.... The skater exerts an action force on the basketball by throwing it. The thrown ball exerts an equal reaction force on the skater, but in the opposite direction. The result is that the basketball moves in one direction and the skater moves in the other direction. *page F43*	**CREATE MENTAL IMAGES** Think about how an artist might draw a picture of what a passage describes. If you can picture what each part of a process looks like, you can tell if you are understanding what you are reading.	"First, I'm picturing an in-line skater. Now I see that she passes a basketball, hard, to someone standing in front of her. The ball goes forward, pushed by her arms and wrists. That's the action force. At the same time, she is pushed backward by the force of the pass. That's the reaction force."

READING STRATEGIES

When reading a text like this...

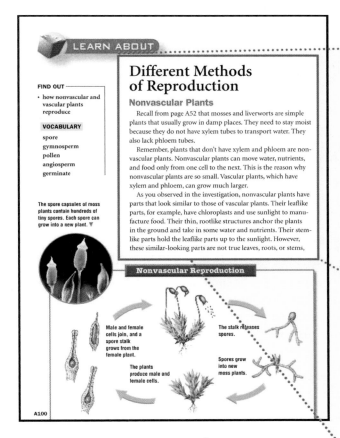

page A100

What you might say to yourself...

"The first heading makes me think that this material is organized to show how plants reproduce. The second heading indicates that the first grouping of plants I'll read about are nonvascular plants.

"As I read this section, I see I'm right. The paragraph tells me how nonvascular plants are different from vascular plants."

A good strategy to use is...

USE TEXT STRUCTURE AND FORMAT

Think about how the information is arranged in the section you are reading. What headings and captions are used? How did the author organize the paragraphs?

- Do the paragraphs tell about causes and effects?
- Do they show the sequence in a process?
- Do they compare two or more things?
- Do they classify concepts?

Recognizing how the information is arranged helps you understand it more easily.

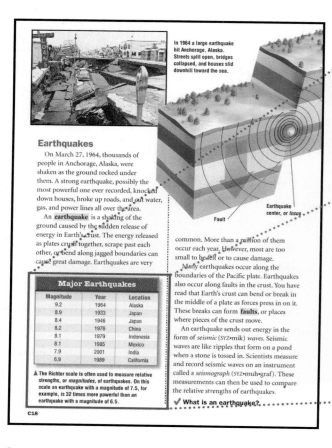

When reading a text like this...

page C18

A good strategy to use is...
USE GRAPHIC AIDS

Look at any pictures, graphs, tables, diagrams, or time lines that are included with the text. Think about why the author included graphic aids.

- Are they there to help you organize the information you are learning? If so, use them as you read.
- Are they there to add new information? If so, take time to read them carefully, to learn the new information.

What you might say to yourself...

"The table and caption accompany text that talks about earthquakes in general. The information here is additional, so I'd better study the table and examine the caption.

"The title of the table tells me what it's about. The headings indicate that the table is comparing the magnitude of major earthquakes. It also tells where and when they occurred. I'm not sure what *magnitude* means, but I notice that it's defined in the caption.

"Now that I know what the table is about, I can start comparing entries. Were any earthquakes the same magnitude? Yes, I see that two were. I also notice that certain areas of the world seem to have powerful earthquakes over and over."

READING STRATEGIES

When reading a text like this...

Electric Charges

You may recall that within an atom, electrons have a negative charge and protons have a positive charge. So the two types of particles attract each other. Most objects have equal numbers of protons and electrons. Sometimes, however, electrons are attracted to the protons of another object and rub off. When an object gains or loses electrons, it has an **electric charge**. An object that has gained electrons has a negative electric charge. An object that has lost electrons has a positive electric charge.

page F68

A good strategy to use is...

REREAD

Think about whether you are understanding what you're reading. If something you read doesn't make sense to you, you may need to go back and read it again. Maybe the passage involves something you learned in an earlier chapter. You may have to look back at that chapter to remember what you learned before.

What you might say to yourself...

"Wait. I'm getting confused. I'd better not read more until I reread this paragraph.

"Now that I've reread it, I see that I need to be clear about the structure of an atom. I'm having trouble remembering exactly what protons and electrons are. I'd better look back at Chapter 2 to remind myself. Then I'll reread this paragraph to see if I understand it."

When reading a text like this...

Why Use Plants to Make Plastic?

Products made from plastics make life simpler, but plastics can cause problems, too. Most plastics are made from petroleum, and when you throw them away, they aren't really gone. Petroleum-based plastics don't decompose (break down) in the environment. Each year people throw away almost 20 million tons of plastic. That's a lot of trash. Now scientists . . .

page A114

A good strategy to use is...

SUMMARIZE AND PARAPHRASE

Think about how you could shorten a passage without leaving out any key ideas.

- Can you sum up the main points of a section in your own words?
- Is there a chapter summary to help you recall the main points?

What you might say to yourself...

"When I summarize this section, I want to include all the important points. I also want to keep it as short as possible. Here is my summary.

"Most plastics are petroleum-based. They don't decompose, so they harm the environment."

Building a Science Vocabulary

Reading and understanding a new science word can be a challenge. One way to make things easier is to use what you already know. First, look at the words that appear near the new word in a sentence or paragraph. Can any of them give you a clue? Next, look at the new science word itself. Have you seen that word—or part of it—before? If so, how was the word used? Using what you already know can help you to learn and remember the meaning of new science words.

Use the Context

Look for Clues in a Sentence or Paragraph

When you see a science word for the first time, look at words that appear near it in a sentence or paragraph. Often, they will provide clues about the meaning of the new word.

For example, look at the vocabulary word *precipitation* in the following sentence:

> Seventy inches of rain and sixty-five inches of snow helped Johnson County set a record for annual **precipitation**.

The sentence refers to rain and snow. You know what both of those things are. So you might be able to figure out that *precipitation* means "water that falls from the clouds."

Look Both Ways

Context clues don't always come before a vocabulary word. Sometimes, clues come after the word. When you see a science word for the first time, don't stop there. Keep reading, and keep your eyes open for context clues.

How could you figure out the meaning of the vocabulary word *submersible* in the following passage?

> To learn more about life in the North Atlantic, scientists will use a **submersible**. Safe inside this underwater vehicle, researchers can look closely at fish and other undersea animals.

If you read beyond the vocabulary word, you will find out that a *submersible* is an underwater vehicle.

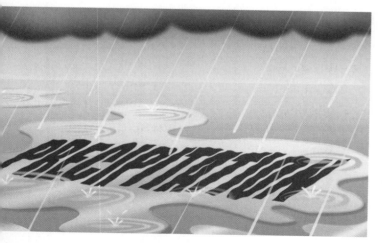

SCIENCE VOCABULARY

Same Words, New Uses

Scientists often use familiar words to describe new things or ideas.

- The word *current*, for example, has long been used to describe flowing water. When scientists needed a word to describe the amount of electricity flowing through a circuit, they chose *current*.

- As science vocabulary words, *crust* and *core* might be unfamiliar to you. But you have used both words before. You know that the crust is the outer surface of a loaf of bread. You also know that the core is the center of an apple.

So, which part of the earth is the crust? You guessed it—the outer surface. And which part is the core? Right—the center!

Other science terms that use familiar words:
greenhouse effect
continental drift
telescope lens
space shuttle

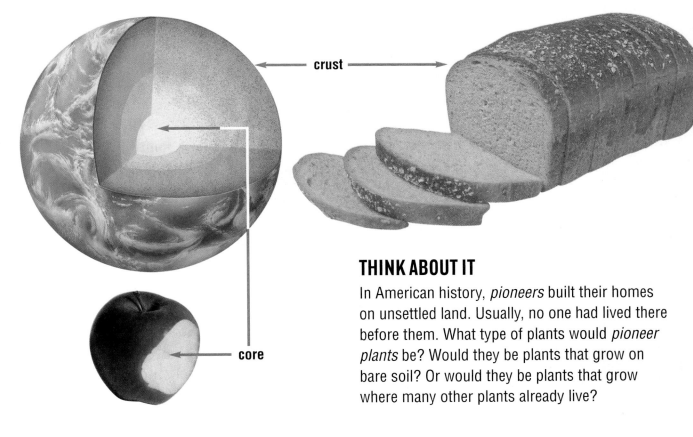

crust

core

THINK ABOUT IT

In American history, *pioneers* built their homes on unsettled land. Usually, no one had lived there before them. What type of plants would *pioneer plants* be? Would they be plants that grow on bare soil? Or would they be plants that grow where many other plants already live?

R45

Greek and Latin Roots in Scientific Terms

Many scientific words in English come from ancient Greek and Latin—two languages used long ago. Scientists often use Greek and Latin roots, or word parts, to make new words.

Consider the word *photosynthesis*. It is used to describe the process by which plants use light to make food for themselves. It comes from the Greek root *photo*, which means "light."

Knowing a word's root can help you figure out its meaning. It can also help you to read long words. Once you recognize a word part like *photo*, there's a better chance that you'll be able to pronounce and understand the remaining part of the word. Other scientific words that use the root *photo* include *photography*, *phototropism*, *photosphere*, and *photometer*.

Photosynthesis

Sunlight is taken in by chlorophyll. Light energy is turned into chemical energy, which is used to split water into hydrogen and oxygen. In a series of reactions, hydrogen combines with the carbon dioxide to form glucose and oxygen.

- carbon dioxide and water
- sunlight and chlorophyll
- glucose and oxygen

Latin and Greek Roots in Science Vocabulary

Root	Meaning	Vocabulary Term(s)	Other Scientific Terms
bios	life	biomass, biome, symbiosis	antibiotic, biochemistry, biology, biosphere
cycle	circle	carbon-oxygen cycle, life cycle, nitrogen cycle, recycling, water cycle	cyclotron, cyclone, cyclometer
derma	skin	epidermis	dermal, dermatology
geo	earth	geothermal, Pangea	geology, geography, geomagnetism
hydro	water	hydroelectric energy	dehydrate, hydraulic, hydrofoil, hydrodynamics
vas	duct	vascular, nonvascular	blood vessel, cardiovascular, vascular bundle

SCIENCE VOCABULARY

Other Common Roots in Scientific Words

Root	Meaning	Example Words
aero	air	aerobic, aerodynamics, aerometer, aeronautics, aerospace
dec	ten	decahedron, decimal
genes	born	genetic, genome
log	study of	biology, geology, physiology, zoology
meso	middle	mesoderm, mesosphere, Mesozoic
pod	foot	arthropod, podiatry, pseudopod
spirare	to breathe	respire, respiratory, transpiration
trope	turn	tropism, phototropism

THINK ABOUT IT

Imagine that you have been hired to name a new invention. The invention is a system of pipes that will carry water to farmers' fields. Which of the following names would be best for the invention?

- Dermal Hydrator
- Hydrocyclon
- Hydrovasculator

Can you use Greek and Latin roots to come up with an even better name? Try using roots to come up with names for imaginary inventions of your own!

Phonics: Sounds and Spellings in Scientific Terms

When you encounter an unfamiliar science word, you need to be able to read and pronounce it correctly. Here are some pointers to help you out.

Some letters and letter combinations always sound the same. In English, some letters and letter combinations almost always stand for the same sound each time they appear.

- The letter combination *th* almost always stands for the sound you hear at the beginning of *thick*. You can find the *th* letter combination in the following vocabulary terms: a**th**racite, ear**th**quake, geo**th**ermal energy, photosyn**th**esis, **th**reatened, and wea**th**ering.

- The letter combination *ph* almost always stands for the same sound as the letter *f*. You can find this letter combination in the following vocabulary terms: atmos**ph**ere, chloro**ph**yll, metamor**ph**osis, ne**ph**rons, **ph**loem, **ph**otosphere, **ph**otosynthesis, **ph**ototropism, and **ph**ysical properties.

Silent *e* makes a vowel long. In English, a silent *e* at the end of a syllable usually means that the vowel that comes before it is a long sound. That means that the vowel "says its own name." You can actually hear the letter name when you say a word with a long vowel. The word *plate*, for example, ends in silent *e*. This means that the *a* that comes before it is a long *a*. Do you hear the *a* when you say *plate*? You can use the silent *e* rule to help you pronounce science words such as gen*o*me (long *o*), sp*o*re (long *o*), photosph*e*re (long *e*), precipit*a*te (long *a*), and bi*o*me (long *o*).

A letter can stand for different sounds, depending on the letter that follows it.

- When the letter *g* is followed by the vowels *a*, *o*, or *u*, it usually sounds like the *g* in *got*. It also sounds this way when it is followed by a consonant. When the letter *g* is followed by *e*, *i*, or *y*, it usually sounds like the *g* in *general*.

  Use this rule to pronounce the *g* sound in the following vocabulary terms: *energy, galaxy, gas, gene, germinate, global warming, grafting, gymnosperm, ligament,* and *magma*.

- When the letter *c* is followed by the vowels *a*, *o*, or *u*, it sounds like the *k* sound in *can*. The letter *c* sounds the same way when it is followed by most consonants. When the letter *c* is followed by the letters *e*, *i*, or *y*, it sounds like the *s* sound in *nice*.

  Use this rule to pronounce the *c* sound in the following vocabulary terms: *acid rain, action force, cell, climate, competition, core, eclipse, ecosystem, extinct, friction,* and *precipitation*.

THINK ABOUT IT

Consider the vocabulary word *recycle*. Are the two *c*'s in the word pronounced in the same way? Why or why not?

SCIENCE VOCABULARY

Rule-breakers can be a challenge. Some letters and letter combinations don't seem to obey any rules at all.

Take the letter combination *ch*, for example. Most of the time, you pronounce it like the *ch* in *chin*. But in many cases, *ch* stands for the *k* sound you hear in the word *echo*. And in words such as *chute*, *ch* can even sound like the *sh* in *ship*.

Use the glossary in your book to help you pronounce these vocabulary terms: *chemical bonds, chlorophyll, chromosome, electric charge, machine, niche,* and *pitch*.

When in doubt, use a glossary or a dictionary. A glossary or a dictionary is the best resource for learning how to pronounce a new science word. Each word in a dictionary appears next to its pronunciation. A guide to the symbols used in the pronunciation usually appears at the front of a dictionary. When you find a new science word and can't figure out how to pronounce it, go straight to a dictionary.

Visit the Multimedia Science Glossary to see illustrations of these words and to hear them pronounced.
www.harcourtschool.com/science

Glossary

As you read your science book, you will notice that new or unfamiliar words have been respelled to help you pronounce them quickly while you are reading. Those respellings are *phonetic respellings*. In this Glossary you will see a different kind of respelling. Here, diacritical marks are used, as they are used in dictionaries. *Diacritical respellings* provide a more precise pronunciation of the word.

When you see the ′ mark after a syllable, pronounce that syllable with more force than the other syllables. The page number at the end of the definition tells where to find the word in your book. The boldfaced letters in the examples in the Pronunciation Key that follows show how these letters are pronounced in the respellings after each glossary word.

PRONUNCIATION KEY

a	**a**dd, m**a**p	m	**m**ove, see**m**	u	**u**p, d**o**ne		
ā	**a**ce, r**a**te	n	**n**ice, ti**n**	û(r)	b**ur**n, t**er**m		
â(r)	c**a**re, **ai**r	ng	ri**ng**, so**ng**	yōō	f**u**se, f**ew**		
ä	p**a**lm, f**a**ther	o	**o**dd, h**o**t	v	**v**ain, e**v**e		
b	**b**at, ru**b**	ō	**o**pen, s**o**	w	**w**in, a**w**ay		
ch	**ch**eck, cat**ch**	ô	**o**rder, j**aw**	y	**y**et, **y**earn		
d	**d**og, ro**d**	oi	**oi**l, b**oy**	z	**z**est, mu**s**e		
e	**e**nd, p**e**t	ou	p**ou**t, n**ow**	zh	vi**s**ion, plea**s**ure		
ē	**e**qual, tr**ee**	ŏŏ	t**oo**k, f**u**ll	ə	the schwa, an unstressed vowel representing the sound spelled		
f	**f**it, hal**f**	ōō	p**oo**l, f**oo**d				
g	**g**o, lo**g**	p	**p**it, sto**p**				
h	**h**ope, **h**ate	r	**r**un, poo**r**				
i	**i**t, g**i**ve	s	**s**ee, pa**ss**		*a* in **a**bove		
ī	**i**ce, wr**i**te	sh	**s**ure, ru**sh**		*e* in sick**e**n		
j	**j**oy, le**dge**	t	**t**alk, si**t**		*i* in poss**i**ble		
k	**c**ool, ta**k**e	th	**th**in, bo**th**		*o* in mel**o**n		
l	**l**ook, ru**l**e	<u>th</u>	**th**is, ba**th**e		*u* in circ**u**s		

Other symbols:
 • separates words into syllables
 ′ indicates heavier stress on a syllable
 ′ indicates light stress on a syllable

R50

GLOSSARY

Multimedia Science Glossary: www.harcourtschool.com/science

acceleration [ak•sel′ər•ā′shən] A change in motion caused by unbalanced forces or a change in velocity **(F35)**

acid rain [as′id rān′] Precipitation resulting from pollution condensing into clouds and falling to Earth **(B99)**

action force [ak′shən fôrs′] The first force in the third law of motion **(F43)**

air mass [âr′ mas′] A large body of air that has nearly the same temperature and humidity throughout **(C75)**

air pressure [âr′ presh′ər] The weight of air **(C65)**

alveoli [al•vē′ə•lē] Tiny air sacs located at the ends of bronchi in the lungs **(A18)**

amphibians [am•fib′ē•ənz] Animals that have moist skin and no scales **(A44)**

angiosperm [an′jē•ō•spûrm′] A flowering plant **(A103)**

asexual reproduction [ā•sek′shoo•əl rē′prə•duk′shən] Reproduction by simple cell division **(A67)**

asteroids [as′tə•roidz] Chunks of rock that look like giant potatoes in space **(D17)**

atmosphere [at′məs•fir] The layer of air that surrounds Earth **(C64)**

atom [at′əm] The smallest unit of an element that has all the properties of that element **(E40)**

axis [ak′sis] An imaginary line that passes through Earth's center and its North and South Poles **(D7)**

B

balanced forces [bal′ənst fôrs′əz] The forces acting on an object that are equal in size and opposite in direction, canceling each other out **(F12)**

biomass [bī′ō•mas′] Organic matter, such as wood, that is living or was recently alive **(F110)**

biome [bī′ōm′] A large-scale ecosystem **(B64)**

birds [bûrdz] Vertebrates with feathers **(A45)**

bone marrow [bōn′ mar′ō] A connective tissue that produces red and white blood cells **(A24)**

capillaries [kap′ə•ler′ēz] The smallest blood vessels **(A17)**

carbon–oxygen cycle [kär′bən ok′sə•jən sī′kəl] The process by which carbon and oxygen cycle among plants, animals, and the environment **(B8)**

cell [sel] The basic unit of structure and function of all living things **(A6)**

cell membrane [sel′ mem′brān′] The thin covering that encloses a cell and holds its parts together **(A8)**

chemical bonds [kem′i•kəl bondz′] The forces that join atoms to each other **(F98)**

chlorophyll [klôr′ə•fil′] A pigment, or coloring matter, that helps plants use light energy to produce sugars **(A96)**

chromosome [krō′mə•sōm′] A threadlike strand of DNA inside the nucleus **(A65)**

R51

classification [klas′ə•fə•kā′shən] The grouping of things by using a set of rules **(A38)**

climate [klī′mit] The average of all weather conditions through all seasons over a period of time **(C80)**

climate zone [klī′mit zōn′] A region throughout which yearly patterns of temperature, rainfall, and amount of sunlight are similar **(B64)**

climax community [klī′maks′ kə•myōō′nə•tē] The last stage of succession **(B93)**

combustibility [kəm•bus′tə•bil′ə•tē] The chemical property of being able to burn **(E24)**

comets [kom′its] Balls of ice and rock that circle the sun from two regions beyond the orbit of Pluto **(D16)**

community [kə•myōō′nə•tē] All the populations of organisms living together in an environment **(B28)**

competition [kom′pə•tish′ən] The contest among organisms for the limited resources of an ecosystem **(B42)**

compound [kom′pound] A substance made of the atoms of two or more different elements **(E48)**

condensation [kon′dən•sā′shən] The process by which a gas changes back into a liquid **(B14, C67, E17)**

conduction [kən•duk′shən] The direct transfer of heat between objects that touch **(F85)**

conductor [kən•duk′tər] A material that conducts electrons easily **(F70)**

conserving [kən•sûrv′ing] The saving or protecting of resources **(B104)**

consumer [kən•sōō′mər] An organism in a community that must eat to get the energy it needs **(B34)**

continental drift [kon′tə•nen′təl drift′] A theory of how Earth's continents move over its surface **(C22)**

convection [kən•vek′shən] The transfer of heat as a result of the mixing of a liquid or a gas **(F85)**

core [kôr] The center of the Earth **(C14)**

corona [kə•rō′nə] The sun's atmosphere **(D41)**

crust [krust] The thin, outer layer of Earth **(C14)**

current [kûr′ənt] A stream of water that flows like a river through the ocean **(C104)**

cytoplasm [sīt′ō•plaz′əm] A jellylike substance containing many chemicals that keep a cell functioning **(A9)**

decomposer [dē′kəm•pōz′ər] Consumer that breaks down the tissues of dead organisms **(B35)**

density [den′sə•tē] The concentration of matter in an object **(E9)**

deposition [dep′ə•zish′ən] The process of dropping, or depositing, sediment in a new location **(C7)**

desalination [dē•sal′ə•nā′shən] The process of removing salt from sea water **(C120)**

diffusion [di•fyōō′zhən] The process by which many materials move in and out of cells **(A10)**

direct development [də•rekt′ di•vel′əp•mənt] A kind of growth where organisms keep the same body features as they grow larger **(A72)**

dominant trait [dom′ə•nənt trāt′] A strong trait **(A79)**

Multimedia Science Glossary: www.harcourtschool.com/science

GLOSSARY

earthquake [ûrth′kwāk′] A shaking of the ground caused by the sudden release of energy in Earth's crust **(C18)**

eclipse [i•klips′] The passing of one object through the shadow of another **(D8)**

ecosystem [ek′ō•sis′təm] A community and its physical environment together **(B28)**

electric charge [i•lek′trik chärj′] The charge obtained by an object when it gains or loses electrons **(F68)**

electric circuit [i•lek′trik sûr′kit] The path along which electrons can flow **(F71)**

electric current [i•lek′trik kûr′ənt] The flow of electrons from negatively charged objects to positively charged objects **(F69)**

electric force [i•lek′trik fôrs′] The attraction or repulsion of objects due to their charges **(F69)**

electromagnet [i•lek′trō•mag′nit] A temporary magnet made by passing electric current through a wire coiled around an iron bar **(F72)**

electron [ē•lek′tron′] A subatomic particle with a negative charge **(E39)**

element [el′ə•mənt] A substance made up of only one kind of atom **(E40)**

El Niño [el nēn′yō] A short-term climate change that occurs every two to ten years **(C83)**

endangered [en•dān′jərd] A term describing a population of organisms that is likely to become extinct if steps are not taken to save it **(B51)**

energy [en′ər•jē] The ability to cause changes in matter **(F62)**

energy pyramid [en′ər•jē pir′ə•mid] Shows the amount of energy available to pass from one level of a food chain to the next **(B38)**

equinox [ē′kwi•noks] Point in Earth's orbit at which the hours of daylight and darkness are equal **(D15)**

erosion [i•rō′zhən] The process of moving sediment from one place to another **(C7)**

estuary [es′chōō•er′ē] The place where a freshwater river empties into an ocean **(B80)**

evaporation [ē•vap′ə•rā′shən] The process by which a liquid changes into a gas **(B14, C67, E16)**

exotic [ig•zot′•ik] An imported or nonnative organism **(B50)**

extinct [ik•stingkt′] No longer in existence; describes a species when the last individual of a population dies and that organism is gone forever **(B51)**

fault [fôlt] A break or place where pieces of Earth's crust move **(C18)**

fiber [fī′bər] Any material that can be separated into threads **(A112)**

fish [fish] Vertebrates that live their entire lives in water **(A44)**

food chain [fōōd′ chān′] The ways in which the organisms in an ecosystem interact with one another according to what they eat **(B35)**

food web [fōōd′ web′] Shows the interactions among many different food chains in a single ecosystem **(B36)**

R53

force [fôrs] A push or pull that causes an object to move, stop, or change direction **(F6)**

fossil [fos'əl] The remains or traces of past life found in sedimentary rock **(C23)**

friction [frik'shən] A force that opposes, or acts against, motion when two surfaces rub against each other **(F6)**

front [frunt] The boundary between air masses **(C75)**

fungi [fun'jī'] Living things that look like plants but cannot make their own food; example, mushrooms **(A39)**

fusion energy [fyōō'zhən en'ər•jē] The energy released when the nuclei of two atoms are forced together to form a larger nucleus **(F112)**

galaxy [gal'ək•sē] A group of stars, gas, and dust **(D54)**

gas [gas] The state of matter that does not have a definite shape or volume **(E14)**

gene [jēn] Structures on a chromosome that contain the DNA code for a trait an organism inherits **(A80)**

genus [jē'nəs] The second-smallest name grouping used in classification **(A40)**

germinate [jûr'mi•nāt] the sprouting of a seed **(A105)**

geothermal energy [jē'ō•thûr'məl en'ər•jē] Heat from inside the Earth **(F111)**

global warming [glō'bəl wôrm'ing] The hypothesized rise in Earth's average temperature from excess carbon dioxide **(C84)**

grain [grān] The seed of certain grasses **(A110)**

gravitation [grav'i•tā'shən] The force that pulls all objects in the universe toward one another **(F8)**

greenhouse effect [grēn'hous' i•fekt'] Process by which the Earth's atmosphere absorbs heat **(C84)**

gymnosperm [jim'nə•spûrm'] Plant with unprotected seeds; conifer or cone-bearing plant **(A102)**

habitat [hab'ə•tat'] A place in an ecosystem where a population lives **(B29)**

hardness [härd'nis] A mineral's ability to resist being scratched **(C37)**

headland [hed'land'] A hard, rocky point of land left when softer rock is washed away by the sea **(C111)**

heat [hēt] The transfer of thermal energy from one substance to another **(F84)**

humidity [hyōō•mid'ə•tē] A measure of the amount of water in the air **(C65)**

hydroelectric energy [hī'drō•ē•lek'trik en'ər•jē] Electricity generated from the force of moving water **(F104)**

igneous rock [ig'nē•əs rok'] A type of rock that forms when melted rock hardens **(C42)**

individual [in'də•vij'ōō•əl] A single organism in an environment **(B28)**

inertia [in•ûr'shə] The property of matter that keeps it moving in a straight line or keeps it at rest **(F41)**

inherited trait [in•her'it•əd trāt'] A characteristic that is passed from parent to offspring **(A78)**

instinct [in'stingkt] A behavior that an organism inherits **(B46)**

insulator [in'sə•lāt'ər] A material that does not carry electrons **(F71)**

Multimedia Science Glossary: www.harcourtschool.com/science

GLOSSARY

intertidal zone [in′tər•tīd′əl zōn′] An area where the tide and churning waves provide a constant supply of oxygen and nutrients to living organisms **(B77)**

invertebrates [in•vûr′tə•brits] Animals without a backbone **(A45)**

jetty [jet′ē] A wall-like structure made of rocks that sticks out into the ocean **(C112)**

joint [joint] A place where bones meet and are attached to each other and to muscles **(A24)**

kinetic energy [ki•net′ik en′ər•jē] The energy of motion, or energy in use **(F62)**

kingdom [king′dəm] The largest group into which living things can be classified **(A39)**

landform [land′fôrm′] A physical feature on Earth's surface **(C6)**

law of universal gravitation [lô′ uv yōōn′ə•vûr′səl grav′i•tā′shən] Law that states that all objects in the universe are attracted to all other objects **(F49)**

learned behavior [lûrnd′ bē•hāv′yər] A behavior an animal learns from its parents **(B46)**

lens [lenz] A piece of clear material that bends, or refracts, light rays passing through it **(F77)**

ligament [lig′ə•mənt] One of the bands of connective tissue that hold a skeleton together **(A25)**

life cycle [līf′ sīkəl] all the stages of an organism's life **(A72)**

light-year [līt′yir′] The distance light travels in one Earth year; about 9.5 trillion km **(D55)**

liquid [lik′wid] The state of matter that has a definite volume but no definite shape **(E14)**

local winds [lō′kəl windz′] The winds dependent upon local changes in temperature **(C73)**

luster [lus′tər] The way the surface of a mineral reflects light **(C37)**

machine [mə•shēn′] Something that makes work easier by changing the size or the direction of a force **(F20)**

magma [mag′mə] Molten rock from Earth's mantle **(C16)**

magnetism [mag′nə•tiz′əm] The force of attraction between magnets and magnetic objects **(F7)**

magnitude [mag′nə•tōōd] Brightness of stars **(D46)**

main sequence [mān′ sē′kwəns] A band of stars that includes most stars of average color, size, magnitude, and temperature **(D47)**

mammals [mam′əlz] Animals that have hair and produce milk for their young **(A44)**

mantle [man′təl] The layer of rock beneath Earth's crust **(C14)**

mass [mas] The amount of matter in an object **(E7)**

mass movement [mas′ mōōv′mənt] The downhill movement of rock and soil because of gravity **(C9)**

R55

matter [mat′ər] Anything that has mass and takes up space **(E6)**

meiosis [mī•ō′sis] The process that reduces the number of chromosomes in reproductive cells **(A68)**

metamorphic rock [met′ə•môr′fik rok′] A type of rock changed by heat or pressure but not completely melted **(C46)**

metamorphosis [met′ə•môr′fə•sis] A change in the shape or characteristics of an organism's body as it grows and matures **(A73)**

microclimate [mī′krō•klī′mit] The climate of a very small area **(C80)**

mineral [min′ər•əl] A natural, solid material with particles arranged in a repeating pattern **(C36)**

mitosis [mī•tō′sis] The process of cell division **(A65)**

molecule [mol′ə•kyōōl′] A grouping of two or more atoms joined together **(E40)**

moneran [mō•ner′ən] The kingdom of classification for organisms that have only one cell and no nucleus **(A39)**

momentum [mō•men′təm] A measure of how hard it is to slow down or stop an object **(F36)**

near-shore zone [nir′shôr′ zōn′] The area beyond the breaking waves that extends to waters that are about 180 m deep **(B77)**

nephrons [nef′ronz′] Tubes inside the kidneys where urea and water diffuse from the blood **(A20)**

net force [net′ fôrs′] The result of two or more forces acting together on an object **(F14)**

neuron [nŏŏr′on′] A specialized cell that can receive information and transmit it to other cells **(A26)**

neutron [nōō′tron′] A subatomic particle with no charge **(E39)**

niche [nich] The role each population has in its habitat **(B29)**

nitrogen cycle [nī′trə•jən sī′kəl] The cycle in which nitrogen gas is changed into forms of nitrogen that plants can use **(B7)**

nonvascular plants [non•vas′kyə•lər plants] Plants that do not have tubes **(A52)**

nuclear energy [nōō′klē•ər en′ər•jē] The energy released when the nucleus of an atom is split apart **(F110)**

nucleus [nōō′klē•əs] **1** *(cell)* The organelle that controls all of a cell's activities **2** *(atom)* The center of an atom **(A8, E39)**

open-ocean zone [ō′pən•ō′shən zōn′] The area that includes most deep ocean waters; most organisms live near the surface **(B77)**

orbit [ôr′bit] The path one body in space takes as it revolves around another body; such as that of Earth as it revolves around the sun **(D7, F48)**

organ [ôr′gən] Tissues that work together to perform a specific function **(A12)**

osmosis [os•mō′sis] The diffusion of water and dissolved materials through cell membranes **(A10)**

Pangea [pan•jē′ə] A supercontinent containing all of Earth's land that existed about 225 million years ago **(C22)**

periodic table [pir′ē•od′ik tā′bəl] The table of elements in order of increasing atomic number, grouped by similar properties **(E47)**

phloem [flō′em] The tubes that transport food in the vascular plants **(A95)**

photosphere [fōt′ə•sfir′] The visible surface of the sun **(D41)**

photosynthesis [fōt′ō•sin′thə•sis] The process by which plants make food **(A96)**

physical properties [fiz′i•kəl prop′ər•tēz] The characteristics of a substance that can be observed or measured without changing the substance **(E6)**

pioneer plants [pī′ə•nir′ plantz′] The first plants to invade a bare area **(B92)**

pitch [pich] An element of sound determined by the speed at which sound waves move **(F79)**

planets [plan′its] Large, round bodies that revolve around a star **(D16)**

plate [plāt] A rigid block of crust and upper mantle rock **(C15)**

pollen [pol′ən] Flower structures that contain the male reproductive cells **(A102)**

pollution [pə•lōō′shən] Waste products that damage an ecosystem **(B99)**

population [pop•yə•lā′shən] All the individuals of the same kind living in the same environment **(B28)**

position [pə•zish′ən] An object's place, or location **(F34)**

potential energy [pō•ten′shəl en′ər•jē] The energy an object has because of its place or its condition **(F62)**

power [pou′ər] The amount of work done for each unit of time **(F19)**

precipitation [pri•sip′ə•tā′shən] Any form of water that falls from clouds, such as rain or snow **(B15, C65)**

prevailing winds [prē•vāl′ing windz′] The global winds that blow constantly from the same direction **(C73)**

producer [prə•dōōs′ər] An organism that makes its own food **(B34)**

protist [prō′tist] The kingdom of classification for organisms that have only one cell and also have a nucleus, or cell control center **(A39)**

proton [prō′ton′] A subatomic particle with a positive charge **(E39)**

radiation [rā′dē•ā′shən] The transfer of thermal energy by electromagnetic waves **(F85)**

reaction force [rē•ak′shən fôrs′] The force that pushes or pulls back in the third law of motion **(F43)**

reactivity [rē′ak•tiv′ə•tē] The ability of a substance to go through a chemical change **(E23)**

receptors [ri•sep′tərz] Nerve cells that detect conditions in the body's environment **(A26)**

recessive trait [ri•ses′iv trāt′] A weak trait **(A79)**

reclamation [rek′lə•mā′shən] The process of restoring a damaged ecosystem **(B110)**

recycle [rē•sī′kəl] To recover a resource from an item and use the recovered resource to make a new item **(B105)**

reduce [ri•dōōs′] To cut down on the use of resources **(B104)**

reflection [ri•flek′shən] The light energy that bounces off objects **(F76)**

refraction [ri•frak′shən] The bending of light rays when they pass through a substance **(F76)**

reptiles [rep′tīlz] Animals that have dry, scaly skin **(A44)**

resistor [ri•zis′tər] A material that resists the flow of electrons in some way **(F71)**

respiration [res′pə•rā′shən] The process that releases energy from food **(B8)**

reuse [rē′yōōz′] To use items again, sometimes for a different purpose **(B105)**

revolve [ri•volv′] To travel in a closed path around an object such as Earth does as it moves around the sun **(D6)**

rock [rok] A material made up of one or more minerals **(C42)**

rock cycle [rok′ sī′kəl] The slow, never-ending process of rock changes **(C52)**

rotate [rō′tāt] The spinning of Earth on its axis **(D7)**

salinity [sə•lin′ə•tē] Saltiness of the ocean **(C97)**

satellite [sat′ə•līt′] A natural body, like the moon, or an artificial object that orbits another object **(D23)**

scuba [skōō′bə] Underwater breathing equipment; the letters stand for **s**elf-**c**ontained **u**nderwater **b**reathing **a**pparatus **(C117)**

sedimentary rock [sed′ə•men′tər•ē rok′] A type of rock formed by layers of sediments that were squeezed and stuck together over a long time **(C44)**

sexual reproduction [sek′shōō•əl rē′prə•duk′shən] The form of reproduction in which cells from two parents unite to form a zygote **(A68)**

shore [shôr] The area where the ocean and land meet and interact **(C110)**

solar energy [sō′lər en′ər•jē] The energy of sunlight **(F111)**

solar flare [sō′lər flâr′] A brief burst of energy from the sun's photosphere **(D42)**

solar wind [sō′lər wind′] A fast-moving stream of particles thrown into space by solar flares **(D42)**

solid [sol′id] The state of matter that has a definite shape and a definite volume **(E14)**

solstice [sol′stis] Point in Earth's orbit at which the hours of daylight are at their greatest or fewest **(D15)**

solubility [sol′yə•bil′ə•tē] The ability of one substance to be dissolved in another substance **(E10)**

sonar [sō′när′] A device that uses sound waves to determine water depth **(C117)**

space probe [spās′ prōb′] A robot vehicle used to explore deep space **(D24)**

species [spē′shēz] The smallest name grouping used in classification **(A40)**

speed [spēd] A measure of the distance an object moves in a given amount of time **(F35)**

spore [spôr] A single reproductive cell that grows into a new plant **(A101)**

streak [strēk] The color of the powder left behind when you rub a material against a white tile called a streak plate **(C37)**

submersible [sub•mûr′sə•bəl] An underwater vehicle **(C117)**

succession [sək•sesh′ən] A gradual change in an ecosystem, sometimes occurring over hundreds of years **(B92)**

sunspot [sun′spot′] A dark spot on the photosphere of the sun **(D42)**

symbiosis [sim′bē•ō′sis] A long-term relationship between different kinds of organisms **(B45)**

system [sis′təm] Organs that work together to perform a function **(A12)**

Multimedia Science Glossary: www.harcourtschool.com/science

GLOSSARY

telescope [tel′ə•skōp′] An instrument that magnifies distant objects, or makes them appear larger **(D23)**

temperature [tem′pər•ə•chər] The average kinetic energy of all the molecules in an object **(F84)**

tendons [ten′dənz] Tough bands of connective tissue that attach muscles to bones **(A25)**

threatened [thret′ənd] Describes a population of organisms that are likely to become endangered if they are not protected **(B51)**

tidal energy [tīd′əl en′ər•jē] A form of hydroelectric energy that produces electricity from the rising and falling of tides **(F106)**

tide [tīd] The repeated rise and fall in the level of the ocean **(C106)**

tide pool [tīd′ pool′] A pool of sea water found along a rocky shoreline **(C111)**

tissue [tish′oo] Cells that work together to perform a specific function **(A12)**

transpiration [tran′spə•rā′shən] The process in which plants give off water through their stomata **(B15)**

unbalanced forces [un•bal′ənst fôrs′əz] Forces that are not equal **(F13)**

universe [yoon′ə•vûrs′] Everything that exists—planets, stars, dust, gases, and energy **(D54)**

vascular plants [vas′kyə•lər plants] Plants that have tubes **(A50)**

velocity [və•los′ə•tē] An object's speed in a particular direction **(F35)**

vertebrates [vûr′tə•brits] Animals with a backbone **(A44)**

villi [vil′ī] Projections sticking into the small intestine **(A19)**

volcano [vol•kā′nō] A mountain formed by lava and ash **(C16)**

volume [vol′yoom] **1** *(measurement)* The amount of space that an object takes up **2** *(sound)* The loudness of a sound **(E8, F79)**

water cycle [wôt′ər sī′kəl] The cycle in which Earth's water moves through the environment **(B14)**

water pressure [wôt′ər presh′ər] The weight of water pressing on an object **(C97)**

wave [wāv] An up-and-down movement of surface water **(C102)**

weathering [weth′ər•ing] The process of breaking rock into soil, sand, and other tiny pieces **(C7)**

weight [wāt] A measure of the pull of gravity on an object **(E7)**

wetlands [wet′landz′] The water ecosystems that include saltwater marshes, mangrove swamps, and mud flats **(B111)**

work [wûrk] The use of a force to move an object through a distance **(F18)**

xylem [zī′ləm] The tubes that transport water and minerals in vascular plants **(A95)**

R59

Abdominal muscles, R26
Absolute magnitude of stars, D46
Acceleration, F35
Acetabular cup, A83
Acid rain, B99, B117
Action force, F43
Activity pyramid, R12
Adaptations, A31, B43, B85
Adrenal glands, R37
Aeronautical engineer, F53
Agricultural scientists, A114–115
Agriculture, saltwater, C122–123
Agronomist, A115, C123, F114–115
Air, water in (chart), B15
Air mass, C75
Air pressure, C66, C89
Air tubes, R32
Algal blooms, B98
Allen, Joe, F24
Alloys, E43
Alpha International Space Station, D26
al-Razi, E50
Aluminum, E41
Alveoli, A18, R32
Alvin submersible, C117–119
Ammonia, B7
Amphibians, A44
Anemometer, C65
Angiosperms, A103
Animals
  with backbones, A44–45, A57
  without backbones, A45–47
  behavior of, B46
  body color of, B40–41
  cells of, A8–9
  diet of, B57
  kingdom of, A39–40
  life cycles of, A72–76, A106–107
  names for, A54–55
Anvil, F79, R22

Apollo 11, D29
Apollo astronauts, D24–25
Apollo program, D23, D29
Apparent magnitude, D46
Aqua-lung, C117
Aristotle, A54–55, C86, E38, E50, F40
Arizona Fish and Game Commission, B18–19
Armstrong, Neil, D29
Arp, Alissa J., B84
Arteries, A17–18, R30
Arthropods, A45
Asexual reproduction, A67
Ash, volcanic, C16
Asphalt, E28–29
Asteroids, D16–18
Astrolabe, D61
Astronauts, A30, D24–25, D29–30, F24–25, F54
Astronomer, D60
Astrophysicist, D59
Atmosphere, C64
Atmospheric conditions, C65
Atomic number, E40
Atomic theory, E38–39
Atoms, E38–40. *See also* Elements
Atrium, R31
Auditory canals, R22
Auditory nerves, R22
Auto mechanic, F115
Autonomic nervous system, R35
Axis, Earth's, D7

Backbones of animals, A42–47, A57
Bacteria (chart), B7
  in food, R10
Balanced forces, F12
Balances, R4
Bald eagles, B52, B114
Ballard, Robert D., C124
Baptist, Kia K., C28
Barometer, C65
Basalt, C43
Bay of Fundy tides (chart), C106

Beaches, C108–113
Beakers, R5
Behavior, animal, B46
Bell, Alexander Graham, F89
Bennett, Jean M., F90
Biceps, R26
Bicycling, R17
Bile duct, R28
Biomass, F110
Biomes. *See* Land biomes
Birds, A45. *See also* Animals
Bjerknes, Jacob, C86–87
Bladder, A20
Bleeding, first aid for, R21
Blood, R31
Blood vessels
  in circulatory system, A16, R30
  in respiratory system, R32–33
Bohr, Niels, E39
Boiling points, E18
Bone marrow, A24, R36
Bones, A24–25, R24–25, R27
Botanists, A56, A116
Boyko, Elisabeth, C122
Boyko, Hugo, C122
Boyle, Robert, E50
Bracken fern, A56
Brackish water ecosystems, B76, B80–81
Brain, A26, F91, R34
Breastbone, R24
Brock microscopes, R3
Bronchi, A18
Budding, A67
Butterfly Pavilion and Insect Center, CO, A120

Calcite mineral, C36
Calendars, D31
California condors, B52–53
Camouflage, B40–41, B42
Canola oil, F114–115
Canopy, B66–67, B70
Cantu, Eduardo S., A84
Cape Hatteras Lighthouse, C112

INDEX

Capillaries, A17, R30
Carbon-oxygen cycle, B8–11, B21
Carbon dioxide
    levels of (chart), C84
    use by plants, B4–5
Carbonic acid, C9
Cardiac muscles, A25
Carnivores, B35
Carson, Rachel, B114–115
Catalytic converters, B104
Cell membrane, A8–9
Cells
    animal versus plant (chart), A8–9
    diffusion and, A10–11
    discovery of, A6
    division of, A64–66
    plant, A8–9
    red blood, A16
    types of, A7
Cell wall, A9
Center of gravity, F27
Central nervous system, A26
Central Park Zoo, NY, B55
Chalcopyrite, C37
Changes in matter, E22–25
Channel Islands National Park, CA, C111
Chemical bonds, F98
Chemical changes in matter, E22–25
Chemical energy, F86
Chemical fertilizers, B96–97
Chemical formulas, E48
Chemical properties, E20–21
Chemical symbols, E47
Chemical vapor deposition (CVD), C55
Chemistry, E50
Chemists, A114–115, C55, E30
Chimpanzees, B54
Chlorine, E41
Chlorophyll, A96
Chloroplasts, A9, A96
Choking, first aid for, R20
Cholera, A28–29
Chromosomes, A8–9, A65–66, A84–85

Cinder volcanoes (chart), C17
Circulatory system, A16–17, R30–31
Cirrus clouds (chart), C68
Classification, xiv
    animals with backbones as, A44–45
    animals without backbones as, A45–47
    chart, A40
    naming in, A54–55
    plants with tubes as, A50–51
    plants without tubes as, A52
    similarities and differences as, A39–40
Clavicle, R24
Climate
    changes in, C83
    human effects on, C84–85
    latitude and, C81
    world, C82
    See also Weather
Climate zone, B64
Climax communities, B93
Clothing fibers, A112
Clouds, C67–69
Cochlea, F79, R22
***Columbia* space shuttle,** A30, D29
Combustibility, E24
Combustion, B9
Comets, D16–18
Communities of organisms, B28
Comparing, xiv
Competition, B42–43
Complete metamorphosis, A74
Composite volcanoes (chart), C17
Compounds, E48–49
Compression, F79
Concave lenses, F77
Condensation, B13, C67, E17
Conduction, F85
Conductivity of metals, E42
Conductors
    electricity, F70
    heat, F85
    sound, F80
Cone-bearing vascular plants, A102

Conglomerate rocks, C44, C51
Congreve, Sir William, D28
Coniferous trees, A102
Connective tissue, A12, A24. *See also* Skeletal system
Conservation
    of energy, F63
    of matter, E26
    of momentum, F44
    of resources, B104
Conservationists, B20, B116
Conserving, B104
Consumers, B34, B38, B82
Continental drift, C22
Continental glaciers, C8
Continents, C20–21
Control variables, xvii
Convection, F85
Convection zone (chart), D41
Convex lens, F77
Cool-down stretches, R14–15
Core, Earth's, C14
Core, sun's (chart), D41
Coriolis, Gustave, C86–87
Cornea, R22
Corona, D40
Cousteau, Jacques, C117
Cowings, Patricia, F54
Craters, moon's, D4–5, D9–10
Creep, C9
Crossbedding (chart), C45
Crust, Earth's, C14
Crystals, C57
Cumberland Island National Seashore, GA, B120
Cumulonimbus clouds (chart), C68
Cumulus clouds (chart), C68
Currents, C104–105, C111
Customary system of measurement, R6
Cycles
    animal versus plant life, A106–107
    carbon, B8–11, B21
    life, A70–75

R61

natural, B6–11
nitrogen, B7
oxygen, B8–11
rock, C48–49, C52–53
sunspot (chart), D42
water, B13–17, B21, C67
Cytoplasm, A9, A65

Dalton, John, E38, E50–51
DDT, B52, B114–115
Decay, B9
Deciduous forests, B64–65, B67
Decomposers, B35
Deep-ocean currents, C105
Deltas, C7
Deltoid muscles, R26
Democritus, E38
Density, E9
Density column, E31
Deposition, C7
Dermis, R23
Desalination, C120
Deserts, B64–65, B69
Desert zone (chart), C82
Diamonds, C36–38, C54–55
Dictionary, R48–49
Diet of animals, B57
Differences in living things, A39–40
Diffusion, A10–11
Digestive system, R28–29
Direct development, A72
Direction of force, F42
Discovery Center of Science & Technology, PA, E56
Disney's Animal Kingdom, FL, B54
Dispersal of seeds, A104
Displaying data, xv
Distillation, E25
DNA, A65–66, A80, A85
Dominant traits, A79
Doppler radar, C87
Douglas, Marjory Stoneman, B20

Ductility, E42
Dunes, C8

Ear canal, F79
Eardrum, F79, R22
Ears, R22
Earth, D16–18
center of, C12–13
changes in surface of, C22–25
core of, C14
crust of, C15
earthquakes and, C18–19
exploring, from space, C26–27
interior of, C14
landforms of, C6, C10
layers of, C29
volcanoes and, C16–17
Earth Day, B115
Earthquakes, C18–19, R19
Eclipses, D8
E. Coli, A29
Ecological physiologists, B84
Ecologists, B83
Ecosystems
animal behavior in, B46
competition in, B42–43
energy flow in, B34–39
energy pyramid of, B38
extinction in, B51
food chains in, B34–35
food webs in, B36–37
habitats and niches in, B29
human effects on, B98–101
limiting factors in, B30–31
natural succession in, B92–95
population comebacks in, B52
population declines in, B50
resources of, B42–44, B104–107
restoring, B110–113, B117
symbiosis in, B45
water. *See* Water ecosystems
Edison, Thomas, F89
Effort, F18
Electric charge, F68
Electric circuits, F66, F70–71

Electric current, F69–70
Electric energy, F66–73
Electric force, F69
Electromagnetic waves, F78
Electromagnets, F72–73
Electrons, E39
Elements
atomic theory of, E38
atoms in, E38–39
common, E41
compounds of, E48–49
discovery of, E50–51
metals as, E42–43
periodic table of, E46–47, E51, E53
El Niño, C83, C87
Embryos, plant, A105
Emergencies, handling, R18–21
Endangered species, B51
Endangered Species Act of 1973, B115
Endocrine system, R37
Energy
chemical, F86
in ecosystems, B34–39
electric, F68–71
electromagnets and, F72–73
kinetic, F62–63
light, F74–78, F88–89
from magnetars, D58–59
potential, F62–63
saving, F117
sound, F78–80, F88–89
of sun, C72–77, D38–39
thermal, F82–85
Energy pyramid, B38, B57
Engineers, E29, F53, F116
English system of measurement (customary), R6
Environment, B26–27, B54–55, B57. *See also* Ecosystems
Environmental awareness, B114–115
Epidermis, R23
Epithelial tissue, A12
Equinox, D15
Erosion, C7–8
Esophagus, A19, R28–29

INDEX

Estuaries, B80–81, C110
Evaporation, B13, C67, C89, E16
Everglades National Park, FL, B20
Everglades: River of Grass (Douglas), B20
Excretory system, A20–21
Exotic organisms, B50
Expanding stars, D49
Experimenting, xi, xvii
Explorer I, D28
Expressive writing, xxi
Extinct, B51
Extinction, B50–52
Eyeglasses, F88
Eyes, F91, R22

Factors, inherited, A79
Fahrenheit, G. D., C87
Family (chart), A40. *See also* Classification
Farnsworth, Philo, F89
Faults, C18
Femoral head osteotomy, A83
Femur, R24–25
Ferns, A50, A56
Ferris, Roxanne S., A56
Fiber, plant, A112
Fibrous roots, A93
Fibula, R24
Field Museum, Chicago, IL, A56
Find Out statements, using, xviii
Fire safety, R18
First aid, R20–21
Fischbach, Ephraim, F26
Fish, A44, B80. *See also* Animals
Fission, A67
Fixing nitrogen, B7
Flicker, John, B20
Flood plains, C7
Flowering plants, A103–105
Focus, earthquake, C18
Fog, C68
Food chains, B34–37, B82–83
Food Guide Pyramid, A111, R8

Food
  from oceans, C120
  from plants, A110–111
Food safety, R10–11
Food web, B36–37
Force(s)
  acting in pairs, F13
  balanced, F12
  calculating, F19
  friction as, F6–7
  gravity as, F8–9
  laws of motion and, F43
  machines and, F20–22
  magnetism as, F7
  net, F14
  spinning, F55
  unbalanced, F13
  work as, F18
Forests, B10–11
  deciduous, B64–65, B67
  taiga, B64–65, B70
  tropical rain, B64–66
Fossil fuels, F98
Fossils, C23–24
Frame of reference, F34
Free nerve endings, R23
Freezing points, E16, E18
Fresh water (chart), B14
Freshwater ecosystems, B76, B78–79
Friction, F6–7
Frog life cycle (chart), A106
Frogs versus grasshoppers (chart), A46
Fronts, C75
Fuels, B8–9, B115
Fungi, A39
Fusion energy, F112

Gabbro, C43
Galactic clusters, D56–57
Galaxies, D54–57
Galena, C36
Galileo, D22–23, F88
Gallbladder, A19, R28

Gametes, A68–69
Ganglia, R35
Gas(es)
  in air (chart), E10
  boiling points and, E18
  changing states and, E16–17
  defined, E14
  particles in, E15
Gastric juices, A19
Gastrocnemius, R26
Gathering data, xv
Genes, A28–29, A80–81, A84. *See also* Inherited traits
Geneticists, A29, A84
Genus, A40, A55. *See also* Classification
Geologists, D30
  structural, C56
Geosat, C26–27
Geoscientists, C28
Geothermal energy, F111
Germination, seed, A105
Gipson, Mack, Jr., C56
Glaciers, B15, C8, C83
Glasswort, C122–123
Glenn, John, D29
Global warming, B115, C84
Global winds, C74
Glossary, using, R48
Goddard, Robert, D28
Gondwana (chart), C22
Gopher tortoise, A54
Gourdine, Meredith, F116
Graduated cylinders, R5
Grain, A110
Grand Canyon, B114, C23
Granite, C43
Grant, Ulysses S., B115
Graphite, C38
Grasshoppers versus frogs (chart), A46
Grasslands, B64–65, B68
Gravitation
  laws of motion and, F40
  universal, F49

R63

Gravity
   center of, F27
   definition of, F8–9
   imaging of, C26–27
   law of universal gravitation and, F49
   micro, F24–25
Gray Herbarium, MA, A56
Great blue heron, B18
Greenhouse effect, C84
Groin muscles, R26
Groundwater, B15, C67
Grouping elements, E44–45
Growth
   cell division as, A64–65
   inherited traits and, A76–81
   life cycles and, A70–75
   regeneration in, A66–67
   sexual reproduction as, A68
Growth rings (chart), A95
Gulf Stream, C104
Gymnosperms, A102
Gypsum, C38

Habitats, B29, B48–49, B51. *See also* Land biomes
Hair cells, F79
Halite, C38
Halley, Sir Edmund, C116
Halophytes, C123
Hammer, ear, F79, R22
Hamstring, R26
Hand lens, R2
Hardness, C37
Harris, Bernard A. Jr., A30
Hawaiian Islands, C17
Headlands, C111
Health Handbook, R8–37
Hearing, R22
Heart
   circulatory system and, A16, R30–31
   nervous system and, R35
   physical activity and, R16–17
   respiratory system and, R33
Heartwood, A51

Heat, C72, F82–85
Heat island, C84
Heavy-machine operator, B19
Hematite, C38
Henry, Joseph, C86
Herbivores, B35
Heredity. *See* Inherited traits
Hertzsprung, Ejnar, D47
Highlands, moon's, D9
Highway engineer, E29
Hip dysplasia, A82–83
Hooke, Robert, A6
Hot spots, volcanic, C26–27
Hubble Space Telescope, D24
Human body systems
   bones and joints, A24–25, R24–25, R27
   cells and, A16
   circulatory, A17, R30–31
   definition of, A12–13
   digestive, A19, R28–29
   excretory, A20–21
   immune, R36
   muscular, A22–23, A25, A46, R26–27
   nervous, A26, R34–35
   respiratory, A18, R32–33
Humerus, R24
Humidity, C65, C67–69
Huntsville-Madison County Botanical Garden, AL, A120
Hybrid versus purebred traits, A80
Hydroelectric energy, F104
Hydrologist, C123
Hygrometer, C65
HyperSoar aircraft, F52–53
Hypothesis, xi–xii, xvi

Ice, landforms changed by, C8
Ice Age (chart), C83
Ice caps (chart), B15
Identifying variables, xvii
Igneous rocks, C42–43, C52
Immune system, R36
Inclined plane, F21

Incomplete metamorphosis, A73
Individual organisms, B28
Industrial Revolution, B10–11, B115
Inertia, F41, F46–47
Inferring, xvi
Informative writing, xxi
Inherited traits
   genes and, A80–81
   Mendel's hypothesis of, A79–81
In-line skating, R16
Inner ear, R22
Instincts, B46
Insulators
   electricity, E42, F71
   heat, F85, F117
   sound, F80
International Chemical Congress, E51
International Space Station, Alpha, D26
Interpreting data, xv, xxiii
Intertidal zone, B76–77
Intestines, R28
Invertebrates, A45–47
Iris, of eye, R22
Iron, E41

Janssen, Hans, F88
Janssen, Zacharias, F88
Jetty, C112
John C. Stennis Space Center, MS, D64
Johnson, Duane, F114–115
Johnson Space Center, F24
Joints, A24–25, R25
Jumping, A46
Jupiter, D16–18

Keck telescope, D50
Kennedy, John F., D28
Key, Francis Scott, D28
Kidneys, A20

INDEX

Kīlauea, C10, C17
Kinetic energy, F62–63
Kingdoms, A39, A55. *See also* Classification
Kloth, Carolyn, C88
Kneecap, R24
Kooyman, Shirley Mah, A116
Krause's endings, R23
Kure Atoll, C17

Lakes (chart), B15
Land biomes
    comparisons of, B72–73
    deciduous forests as, B67
    deserts as, B69
    grasslands as, B68
    taiga as, B70
    tropical rain forests as, B66
    tundra as, B71
Landfills, B102–103, B106
Landforms
    changing, C6, C10, C29
    effects of ice on, C8
    oceanic, C26–27
    See also Earth
Landslides, C9
Large intestine, R28
Larva, A74
Latitude, C81
Lava, C16
Law of conservation of matter, E26
Law of universal gravitation, F49
Laws of motion, F24, F40–43, F55
Leapheart, Theophilus, E30
Learned behavior, B46
Leaves, A57, A92, A96, A117
Lens(es)
    eyes, F77, R22
    hand, R2
Levers, F20–22
Life cycles
    animal versus plant, A106–107
    description of, A72–75

Ligaments, A25, R25
Light, F74–78, F88–89
Light microscopes, R3
Light-years, D55
Limestone, C44
Limiting factors, B30–31, B43 (chart)
Linnaeus, Carolus, A55
Lippershey, Hans, F88
Liquids, E14–19, E31, R5
Liver, A19, R28
Liverworts, A52
Local environment, B26–27
Local winds, C73
Loma Linda University School of Medicine, CA, A29
Longshore currents, C111
Lowry Park Zoo, FL, B54
Lunar eclipses (chart), D8
Lungs, A18, A31, R16–17, R31–32
Luster, C37, E42
Lymph nodes, R36

Machines, F20–22
Magma, C16
Magnetars, D58–59
Magnetism, D58–59, F4–5, F7, F27, F72–73
Magnetite, C36
Magnitude, D46
Main ideas, xix
Main sequence stars, D47, D49
Malleability, E42
Mammals, A44. *See also* Animals
Mantle, Earth's, C14
Marble, C46
Marconi, Guglielmo, F89
Maria, of moon, D9
Mars, D16–18
Mass, E7, F38–39, F42
Mass movement, C9
Matter
    conservation of, E26
    density of, E9
    mass and weight of, E7

    melting and boiling points of, E16
    mixtures and solutions of, E10–11
    particles of, E15
    physical and chemical changes in, E22–25
    states of, E14, E16–17
    volume of, E8
Maya Indians, D22
McClendon, Dorothy, B56
Mealworms, A70–71
Measurement
    of liquids, R5
    in science experiments, xiv, xxii
    systems of, R6
    of weather conditions, C62–63, C65
    of work, F16–17
Meiosis, A68
Meissner's endings, R23
Meitner, Lise, E52
Melting points, E16, E18
Mendel, Gregor, A79
Mendeleev, Dmitri, E46, E51
Mercury, D12, D17–18
Mercury program, D23
Merkel's endings, R23
Metaconglomerate rocks, C51
Metalloids, E46
Metals, E42–43
Metamorphic rocks, C46, C51, C52
Metamorphosis, A73–74
Meteorologists, C68, C88
Metersticks, R5
Metric system, R6
Mexia, Ynes E. J., A56
Mica, C36
Microbiologist, B56
Microclimates, C80. *See also* Climates
Microgravity, F24–25. *See also* Gravity
Microscopes, A4–6, R3
    scanning tunneling, E51

R65

Middle ear, R22
Milky Way Galaxy, D52–57
Minerals
    evaporation and, C57
    in oceans, C120
    properties of, C36–37
    uses for, C38–39
Mir Space Station, D29
Mission specialists, F25
Mitochondria, A9
Mitosis, A65–67
Mixtures, E10–11, E25, E31
Mohs' hardness scale, C37
Molecules, E40
Mollusks, A45
Molting, A73
Momentum
    conservation of, F44
    definition of, F36
Monerans, A39
Montreal Protocol, B115
Moon
    calendar and, D31
    eclipses and, D8
    exploring, D24–25
    orbit of, F48–51
    in space, D6–7
    surface of, D9
    tides and, C106
Morse, Samuel F. B., F89
Mosses, A52
Motion
    acceleration and, F35
    momentum and, F36–37, F44
    Newton's laws of, F40–43
    position and, F34
    speed of, F35
Mountains, C16
Mountain zone (chart), C82
Mount St. Helens, WA, B94–95, C16
Mouth
    in digestive system, R28
    in respiratory system, R32
Mucous membranes, nose, R23
Mudslides, C9
Mukai, Chiaki, F24
Muscular system, A22–23, A25, A46, R26–27

Musical instruments, F88
Mutualism, B45

Narrative writing, xxi
Nasal bones, R23
Nasal cavity, R23
National Aeronautics and Space Administration (NASA), D28
National Air and Space Museum, Washington, D.C., D64
National Audubon Society, B20
National Science Center, GA, F120
National Weather Service (NWS), C86, C128
National Zoo, Washington D. C., B55
Natural cycles, B6–11
Natural History and Antiquities of Selborne, The (White), B114
Natural succession, B92–95
Nautile, C118–119
Navarro, Julio, D60
Near-shore zone, B76, B77
Nebulae, D48, D56–57
Nephrons, A20
Neptune, D16–18
Nerves, R34
Nervous system, A12, A26, R34–35
Net force, F14
Neurons, A26–27, R34
Neutral atoms, E39
Neutral Buoyancy Laboratory (NBL), F24–25
Neutrons, E39, E51
Newton, Sir Isaac, D22–23, F40–43, F49, F51
Niches, B29
Nitrates, B7
Nitrogen cycle, B7
NMR spectroscopy, E30
Nodules, plant roots (chart), B7
Nonvascular plants, A52, A98–99, A100–101

Nose, R23, R32
Nostril, R23
Nuclear energy, F110
Nuclear physicist, E52
Nucleus
    of atoms, E39
    of cells, A8–9, A65 (chart)
Numbers, how scientists use, xxii–xxiii
Number sense, xxiii
Nutrients, A17
Nutrition, R8–11

Observing, x, xiv
Obsidian, C43
Oceans
    currents in, C104–105
    exploring, C26–27, C116–119
    food chains in, B82–83
    resources of, C120–121
    shorelines of, C108–113
    tides of, C106–107
    waves in, C102–103
    zones of, B76–77
Oil spills, F117
Olfactory bulb, R23
Olfactory tract, R23
Omnivores, B35
Open-ocean zone, B77
Optic nerves, R22
Orangutans, B55
Orbitals, atomic (chart), E39
Orbits
    of Earth's moon, D7
    of planets, F48–51
Order, A40
Oregon Dunes National Recreation Area, OR, B120
Organelles, A8
Organic chemists, C55
Organisms, communities of, B28
Organs, A12–13. *See also* Human body systems
Osmosis, A10–11
Outer ear, R22

Oxbow lake (chart), C10
Oxygen
   in circulatory system, A17
   cycle, B8–11
   element, E41
   in respiratory system, R33
Ozone layer, B115

P

Pacini's endings, R23
Paired forces, F13
Paleontologists, C25
Pancreas
   in digestive system, A19, R28
   in endocrine system, R37
Pangea, C22
Parallel circuit (chart), F70
Parathyroid glands, R37
Patella, R24
Pea plant life cycle (chart), A106
Pelvis, R24
Peregrine falcons, B52
Periodic table of elements, E46–47, E51, E53
Peripheral nervous system, R35
Permafrost, B71
Persuasive writing, xxi
Pesticides, B98–99
Phalanges, R24
Phloem, A95
Phonics, R48–49
Phonograph, F89
Photosphere, D40, D41
Photosynthesis, A96, B8–9
Phylum (classification chart), A40
Physical activity
   heart and lungs and, R16–17
   planning for, R12
   pyramid of, R12
   stretching in, R14–15
   workout guidelines for, R13
Physical properties, E6
Physicians, A30
Physicists, E52, F26, F90, F116
Physiologists, B84
Phytoplankton, B82–83

Pigment, A96
Pioneer plants, B92–93
Pitch, F79
Pituitary gland, R37
Planaria, A67
Planetary nebulae, D48
Planets, D14–19, F46–51
Planning investigations, x–xii, xvi
Plants
   cells of, A6–9
   classification of, A50–53
   as fiber, A112
   as food, A110–111
   kingdom of, A39–40
   life cycles of, A106–107
   names for, A54–55
   nonvascular, A100–101
   pioneer, B92–93
   products from, A112, A114–115
   reproduction in, A100–105
   restoring ecosystems and, B117
   vascular. *See* Vascular plants
   water and, A57
Plasma, A17
Plastics, A114–115
Platelets, A17
Plates, Earth's, C14–15, C26–27
Pluto, D16–18, D31
Polar bears, B55
Polar zone (chart), C82
Pollen, A102
Pollution, B99
Polylactic acid (PLA) resin, A115
Pond ecosystems, B78–79
Popcorn, A108–109
Population, B28
Population density, B31
Populations of organisms, B28
Position, F34
Potatoes, vaccines from, A28–29
Potential energy, F62–63
Power, F19
Prairie food web, B36–37
Precipitation, B15–16, B30, B72, C65, C67

Predators, B42
Predicting, xv
Prevailing winds, C73–74
Primary succession, B92–93
Process skills, xiii–xvii
Producers, B34, B38, B66, B82
Prop roots, A93
Protists, A39
Protons, E39, E51
Protostars, D49
Psychophysiologists, F54
Pukapuka Ridges, Tahiti, C27
Pulleys, F20
Pulmonary arteries and veins, A18
Pulp, wood, A112
Pumice, C43
Pupa, A74
Pupil, eye, R22
Purebred versus hybrid traits, A80

Q

Quadriceps, R26
Quartzite, C46

R

Radiation, F85
Radiation zone, sun (chart), D41
Radio telescopes, D23
Radius, R24
Rainfall. *See* Land biomes; Precipitation; Weather
Rain forests, tropical, B66
Rain gauge, C65
Rarefaction, F79
Reaction force, F43
Reactions, E53
Reactivity, E23
Reading, how scientists use, xviii–xx
Reading strategies, R38–43

Receptors, A26
Recessive traits, A79
Reclamation, B110
Recording data, xv
Recycle resources, B104, B105
Recycling Act of 1976, B115
Red giants, D47–49
Reduce resource use, B104–105
Reflected light, F74–75
Reflection, F76
Reflexes, A26–27
Refraction, F76
Regeneration, A66–67
Remote Manipulator System (RMS), F25
Reproduction
  asexual, A67
  in cells, A62–63
  in cone-bearing vascular plants, A102
  in flowering vascular plants, A103–105
  life cycles and, A106–107
  in nonvascular plants, A100–101
  sexual, A68
  in simple vascular plants, A101
Reptiles, A44. *See also* Animals
Resistors, F71
Resonance, F80
Resources
  competition for, B42–43
  ocean, C120–121
  sharing, B43–44
  symbiosis and, B45
  using, B104–107
Respiration, B8
Respiratory system, A18, R32–33
Restoring ecosystems, B110–113, B117
Retina, R22
Reuse of natural materials
  balance in, B10–11
  carbon-oxygen cycle in, B8–9
  nitrogen cycle in, B7
Reuse resources, B104–105

Revolution (Earth-moon system), D6–7
Revolve, D6
Rib cage, R24–25
Rip currents, C104
Rivers, B15, B110
Rocket-based combined cycle (RBCC) engine, F52–53
Rockets, D28–29, F55
Rock cycle, C48–49, C52–53
Rocks
  changes in, C50–51
  igneous, C42–43
  metamorphic, C46
  records in, C23. *See also* Earth
  sedimentary, C44–45
  weathering of, C57
Rogue waves, C103
Root hairs, A93
Roots, plant, A92–94
Roots of words, R46–47
Rotate, D7
Rotation, Earth's, D7
Ruffini's endings, R23
Runoff, C67
Russell, Henry, D47
Russell, James T., F89
RXTE satellite, D50

S

Safety, xxiv, R10–11, R18–21
Salinity, C97
Saliva, A19
Salivary glands, R29
Salt water (chart), B14
Saltwater agriculture, C122–123
Saltwater ecosystems, B76, B77
San Andreas fault, CA, C15
Sand spits, C110
Sandstone, C45
Sapwood, A51
Satellite images, C26–27
Satellites, D23, D50
Satellite technician, C27
Saturn, D16–18
Scales, on conifers, A102
Scanning tunneling microscope, E51

Scavengers, B39
Sceptical Chymist, The (Boyle), E50
Schist, C46
Schmitt, Harrison, D30
Science reading strategies, R38–43
Science tools, R2–5
Science vocabulary, R44–49
Scientific methods, x–xii
Scientists, how they work, x–xxiv
Screw, F22
Scuba, C117
Sea-viewing Wide Field Sensor (Sea-WiFS), B82–83
Secondary succession, B94–95
Sediment, C7
Sedimentary rocks, C44–45, C52–53
Seedlings, A105
Seeds, A104–107
Seismic waves, C18
Seismograph, C18
Self-healing asphalt, E28–29
Semicircular canals, R22
Sense organs, R22–23
Series circuit (chart), F70
Serving size, R9
Sexual reproduction, A68
Shellfish, B80
Shield volcanoes (chart), C17
Shoreline currents, C104
Shores, C108–113
Show Low wetland, AZ, B18–19
Sight, R22
Silent Spring (Carson), B114–115
Silicon, E41
Siltstone, C44
Silver, C38
Simple vascular plants, A101
Sinkholes, C9
Skeletal system, A25, A31, A47, R24–25, R27
Skin, A7, R23
Slate, C46
Small intestine, R28
Smell, R23
Smooth muscles, A25

INDEX

Soaps from plants, A112
Sodium, E41
Soil, water in (chart), B15
Solar eclipses (chart), D8
Solar energy, B9, D38–39
Solar flares, D42
Solar system, D22–27, D31
Solar wind, D42
Solids, E14–19
Solstice, D15
Solubility, E10
Solutions, E10–11
Sonar, C117
Sound, F78–81, F88–89, F91
Space, A30, C26–27, D22–27, F52–53
Spaceflight, D28–29
Space probes, D24
Space shuttles, A30, D29
Space stations, D26, D29
Spacesuits, D25
Species, A40, A55. *See also* Classification
Speed, F34–37
Speed of sound (chart), F80
Spinal cord, A26, R34, R35
Spine, R24, R25
Spinning, F55
Spleen
    in circulatory system, R30
    in immune system, R36
Spores, A100–101
Spring scale, R4
Sputnik I, D23, D28
Stars
    changes in, D48–49
    magnetar-type, D58–59
    magnitude of, D46
    navigation by, D61
    observing, D50–51
    types of, D47
"Star-Spangled Banner" (Key), D28
States of matter, E12–19
Static electricity, F69
Stem, plant, A48–49, A92, A94–95
Sternum, R24

Stirrup, ear, F79, R22
Stomach, A19, R28, R29
Stomata, A96, A117
Stopwatch, R5
Storage roots, plant, A94
Storm surges, C103
Stratus clouds (chart), C68
Streak (chart), C37
Stretching, R14–15
Strip mining, B100
Subcutaneous layer, R23
Sublimation, E17
Submersibles, C117–119
Succession, B92–95
Sukumar, Raman, B116
Sun
    energy from, D38–39
    evaporation and, C89
    features of, D42–43
    model of, D61
    stars compared to (chart), D47
    structure of, D40–41
    weather and, C72–77
Sunspots, D36–37, D40, D42
Supergiant stars, D47
Supersonic flight, D28
Surface currents, C104, C105
Swimming, R16
Symbiosis, B45
Synapse, A26
System, A12

T

Taiga, B64–65, B70
Taproots, A93
Taste, R23
Taste buds, R23
Telegraph, F89
Telephone, F89
Telescopes, D22–24, D50–51, F88
Temperate zone (chart), C82
Temperature, F84
    of earth (chart), C84
    ice ages and, C83
Tendons, A25, R25, R27

Tennis, R17
Terminal moraines, C10
Thermal energy, F84–85
Thermometers, C65, R2
Thomas Jefferson National Accelerator Facility, VA, E56
Threatened, B51
Thunderstorm safety, R19
Thymus gland
    in endocrine system, R37
    in immune system, R36
Thyroid gland, R37
Tibia, R24
Tidal energy, F106
Tide pools, C111
Tides, C106–107
Timing devices, R5
Tissues, A12–15
    connective, A24
    See also Human body systems
Titanic, C118–119
Tongues, R23
Tonsils, R36
Tools, science, R2–5
Total hip replacement (THR), A82–83
Touch, R23
Trachea, A18, R32
Trade winds, C87
Traits, A78. *See also* Inherited traits
Transformation of energy, F63
Transpiration, B15
Trapezius, R26
Trees, A51, A95. *See also* Forests; Land biomes; Plants
Tropical rain forests, B64–66
Tropical zone (chart), C82
Tsunamis, C103
Tubes, plant, A50–51. *See also* Vascular plants
Tundra, B64–65, B71
Tycho crater, D9

U.S. Army Tank Automotive Command (TACOM), B56
U.S. Forest Service (USFS), B18–19
Ulna bone, R24
Unbalanced forces, F13
Undersea explorers, C124
Universe, D54
Universal gravitation, F49
Updrafts, C73
Uranus, D16–18
Ureters, A20
Urethra, A20

Vaccines, A28–29
Vacuoles, A9
Valley glaciers, C8
Van de Graaff generator, F66
Vascular plants
    leaves of, A92, A96
    parts of, A50–51
    reproduction in, A101–107
    roots of, A92–94
    stems of, A92, A94–95
Veins, A17–18, R30
Velocity, F35, F38–39
Ventricles, R31
Venus, D16–18
Vertebrae, R25
Vertebrates, A44–45. *See also* Animals
Very Large Array radio telescopes, NM, D50
Veterinary technicians, A83
Viking program, D24
Villi, A19
Volcanoes, C10, C16–17
Volume, E8, F79
Voluntary muscles, A25
Voyager program, D24

Walking, R17
Warm-up stretches, R14–15
War of 1812, D28
Washers, F22
Wastewater, B108–109
Water
    adult human daily use of (chart), A20
    cycle of, C67
    on Earth versus moon, D10
    ecosystems of. *See* Water ecosystems
    humans and, B16–17
    landforms changed by, C7
    in plants, A117
Water cycle, B12–17, B21, C67
Water ecosystems
    estuaries, B80–81
    freshwater, B76, B78–79
    interactions in, B85
    saltwater, B76–77
Water pressure, C97
Water vapor, C67–69
Waves
    electromagnetic, F78
    ocean, C102–103
    sound, F79–80, F91
Weather
    air pressure and, C66
    forecasting, C86–87
    measuring, C65
    occurrence of, C64
    sun and, C72
    water and, C67–69
    winds and, C73–75
    See also Climate; Water cycle
Weathering, C7, C50, D10
Wedge, F20
Weight, E7
Weightlessness, F24–25
Westerlies, C74

Wetlands, B18–19, B110–111
Wheel and axle, F21–22
White, Gilbert, B114–115
White blood cells, R36
White dwarf stars, D48
Windpipe, A18, R32
Wind
    landforms changed by, C8
    prevailing, C73–75
    waves and, C102–103
Wind vane, C65
Wood, A112
Woodchucks, A54
Woolly mammoths, C24
Work
    calculating, F19
    effort in, F18
    machines and, F20–22
    measurement of, F16–17
    weightless, F24–25
Working out, R13
World climates, C82
Wrist, R25
Writing, how scientists use, xxi

Xylem, A95

Yellowstone National Park, B94–95, B115

Zoo guides, B55
Zooplankton, B82–83
Zoos, B54–55
Zworykin, Vladimir, F89
Zygotes
    in reproduction, A68, A101–102

Page Placement Key:
(l)-left, (r)-right, (t)-top, (c)-center, (b)-bottom, (bg)-background, (fg)-foreground, (i)-inset

Cover and Title Pages
Wolfgang Kaehler/Corbis; (bg) Eduardo Garcia/FPG International

Table of Contents
iv (t) Anup & Manoj Shah/Animals Animals; iv (bg) Grant V. Faint/The Image Bank; v (t) Zig Leszczynski/Animals Animals; v (bg) Karl Hentz/The Image Bank; vi (t) Eric & David Hosking/Photo Researchers; vi (bg) Bios (Klein-Hubert)/Peter Arnold, Inc; vii Telegraph Colour Library/FPG International; viii (t) Alvis Upitis/The Image Bank; viii (bg) Tim Crosby/Liaison International; ix (t) David Zaitz/Photonica; ix (bg) Stone;

Unit A
Unit A Opener (fg) Anup & Manoj Shah/Animals Animals; (bg)Grant V. Faint/The Image Bank; A2-A3 Image Shop/Phototake; A3 (l) Lawrence Migdale/Photo Researchers; A3 (c) Quest/Science Photo Library/Photo Researchers; A4 Charles D. Winters/Timeframe Photography, Inc./Photo Researchers; A6 (l) The Granger Collection, New York; A6 (c), (r) Courtesy of Hunt Institute for Botanical Documentation, Carnegie Mellon University, Pittsburgh, PA; A7 (tl) Ed Reschke/Peter Arnold, Inc.; A7 (tr) Michel Viard/Peter Arnold, Inc.; A7 (bl) Courtesy of Dr. Sam Harbo D.V.M., and Dr. Jurgen Schumacher D.V.M. , Veterinary Hospital, University of Tennessee; A7 (br) A.B. Sheldon/Dembinsky Photo Associates; A8 Dwight R. Kuhn; A9 Courtesy of Dr. Sam Harbo D.V.M., and Dr. Jurgen Schumacher D.V.M., Veterinary Hospital, University of Tennessee; A11 Skip Moody/Dembinsky Photo Associates; A14 Michael Newman/PhotoEdit; A16 (l) Dr. Tony Brain/Science Photo Library/Photo Researchers; A16 (r) Prof. P. Motta/Dept. of Anatomy/University "La Sapienza", Rome/Science Photo Library/Photo Researchers; A22 Gary Holscher/Stone; A28 D. Cavagnaro/DRK; A28 (i) Dr. Dennis Kunkel/Phototake; A29 Mark Richards/PhotoEdit; A30 NASA; A33 Charles D. Winters/Timeframe Photography, Inc./Photo Researchers; A34-A35 Gregory Ochocki/Photo Researchers; A35 (t) Dave Watts/Tom Stack & Associates; A35 (b) Frances Fawcett/Cornell University/American Indian Program; A36 Christian Grzimek/Okapia/Photo Reseachers; A38-A39 Bill Lea/Dembinsky Photo Associates; A38 (l) MESZA/Bruce Coleman, Inc.; A38 (c) Andrew Syred/SPL/Photo Researchers; A38 (r) Robert Brons/BPS/Stone; A39 (t) Bill Lea/Dembinsky Photo Associates; A39 (tc) Dr. E. R. Degginger/Color-Pic; A39 (c) S. Nielsen/Bruce Coleman, Inc.; A39 (bc) Robert Brons/BPS/Stone; A39 (b) Andrew Syred/SPL/Photo Researchers; A41 Daniel Cox/Stone; A42 Arthur C. Smith, III/Grant Heilman Photography; A44 (t) Ana Laura Gonzalez/Animals Animals; A44 (b) Tom Brakefield/The Stock Market; A44-A45 Runk/Schoenberger/Grant Heilman Photography; A45 (tl) Amos Nachoum/The Stock Market; A45 (tc) Hans Pfletschinger/Peter Arnold, Inc.; A45 (tr) Mark Moffett/Minden Pictures; A45 (br) Larry Lipsky/DRK; A46 (t) James Balog/Stone; A46 (b) Stephen Dalton/Photo Researchers; A48 Darrell Gulin/Stone; A50 Dr. E. R. Degginger, FPSA/Color-Pic; A51 Phil A. Dotson/Photo Researchers; A52 (t) Heather Angel/Biofotos; A52 (c) Runk Schoenberger/Grant Heilman Photography; A52-A53 Runk Schoenberger/Grant Heilman Photography; A54 Leonard Lee Rue III/Photo Researchers; A54-A55 S. J. Krasemann/Peter Arnold, Inc.; A55 (tl) Art Resource, NY; A55 (tr) Dr. E. R. Degginger/Photo Researchers; A55 (bl) Superstock; A55 (br) The Granger Collection, New York; A56 (t) Courtesy of Hunt Institute for Botanical Documentation, Carnegie Mellon University, Pittsburg, PA; A56 (b) Grant Heilman Photography; A60-A61 Rob & Ann Simpson/Visuals Unlimited; A61 (l) Dwight R. Kuhn; A61 (r) Dr. D. Spector/Peter Arnold, Inc.; A62 Ron Kimball; A64 (l) Jerome Wexler/Photo Researchers; A64 (cl), (c) Carolina Biological Supply Company/Phototake; A64 (cr) Jerome Wexler/Photo Researchers; A64 (r) Kenneth H. Thomas/Photo Researchers; A65 Conly L. Rieder/BPS/Stone; A66 (tl), (tc), (tr) Carolina Biological Supply Company/Phototake; A66 (c) Noble Proctor/Photo Researchers; A66 (b) Zig Leszczynski/Animals Animals; A67 (tl), (tc), (tr) Carolina Biological Supply Company/Phototake; A67 (b) Bob Gossington/Bruce Coleman, Inc.; A69 Carolina Biological Supply Company/Phototake; A70 J.H. Robinson/Photo Researchers; A72 (t) Peter A. Simon/Phototake; A72 (b) Dr. E.R. Degginger/Photo Researchers; A73 (l) Thomas Gulz/Visuals Unlimited; A73 (c) Dwight R. Kuhn; A73 (r) William J. Weber/Visuals Unlimited; A74 Harry Rogers/Photo Researchers; A75 Michael Fogden/Bruce Coleman, Inc.; A76 Paul Barton/The Stock Market; A78 Phil Savoie/The Picture Cube; A79 The Granger Collection, New York; A82 Tim Davis/Tony Stone Images; A83 (li) College of Veterinary Medicine/University of Florida; A83 (r) Zigy Kaluzny/Tony Stone Images; A84 Henry Friedman/HRW; A84 (i) Oliver Meckes/Photo Researchers; A88-A89 Tom Bean/Stone; A89 (t) Inga Spence/Visuals Unlimited; A89 (b) Ned Therrien/Visuals Unlimited; A90 James Randklev/Stone; A92 (l) Richard Choy/Peter Arnold, Inc.; A92 (r) Reinhard Siegel/Stone; A93 Norman Myers/Bruce Coleman, Inc.; A93 (li) Dr. E. R. Degginger/Color-Pic; A93 (ri) John Kaprielian/Photo Researchers; A95 Jane Grushow/Grant Heilman Photography; A96-A97 (t), A96 (ti) Runk/Schoenberger/Grant Heilman Photography; A96-A97 (b) Alan Levenson/Stone; A98 Darrell Gulin/Dembinsky Photo Associates; A100 Kim Taylor/Bruce Coleman, Inc.; A101, A102 (t) Runk/Schoenberger/Grant Heilman Photography; A102 (b) S.J. Krasemann/Peter Arnold, Inc.; A103 (t) Dr. E. R. Degginger/Color-Pic; A103 (b) Robert Maier/Earth Scenes; A104 (t) David Cavagnaro/Peter Arnold, Inc.; A104 (tc) E. R. Degginger/Bruce Coleman, Inc.; A104 (bc) Gregory K. Scott/Photo Researchers; A104 (b) Kevin Schafer Photography; A104 (bg) Jeff Lepore/Photo Researchers; A105 Runk/Schoenberger/Grant Heilman Photography; A106 (animal life cycle) (t) Gregory K. Scott/Photo Researchers; A106 (r) Harry Rogers/National Audubon Society; A106 (b) David M. Dennis/Tom Stack & Associates; A106 (l) Jen & Des Bartlett/Bruce Coleman, Inc.; A106 (plant life cycle) (t) Dr. E. R. Degginger/Color-Pic; A106 (r) Barry L. Runk/Grant Heilman Photography; A106 (b) Jane Grushow/Grant Heilman Photography; A106 (l) Dwight R. Kuhn; A112 (l) Alan & Linda Detrick/Photo Researchers; A112 (cr) Angelina Lax/Photo Researchers; A113 Grant Heilman Photography; A113 (i) Will & Deni MvIntyre/Photo Researchers; A114-115 Dana Downie/AGStock USA; A115 (b) Mark Richards/PhotoEdit; A116 Dennis Carlyle Darling/ HRW; A118 Dr. E. R. Degginger/Color-Pic; A119 James Randklev/Stone; A120 (t) Jeff Greenberg/Unicorn Stock Photos; A120 (b) Jack Olson Photography;

Unit B
Unit B Opener (fg) Zig Leszczynski/Animals Animals; (bg) Karl Hentz/The Image Bank; B2 Clyde H. Smith/Peter Arnold, Inc.; B2-B3 Superstock; B3 Earl Roberge/Photo Researchers; B4 Wolfgang Kaehler Photography; B6 Randy Ury/The Stock Market; B7 Thomas Hovland/Grant Heilman Photography; B10 Wolfgang Kaehler Photography; B12 Michael Giannechini/Photo Researchers; B14-B15 Greg Vaughn/Stone; B16 (t) C. Vincent/Natural Selection Stock Photography; B16 (b) Bob Daemmrich Photography, Inc.; B16 (bi) Superstock; B18 John Shaw/Bruce Coleman, Inc.; B18-B19 Lee Rentz/Bruce Coleman, Inc.; B19 Ken Graham/Bruce Coleman, Inc.; B20 Sipa Press; B22 Greg Vaughn/Stone; B23 C. Vincent/Natural Selection Stock Photography; B24-B25 P & R Hagan/Bruce Coleman, Inc.; B25 (t) Tomas del Amo/Pacific Stock; B25 (b) Mitsuaki Iwago/Minden Pictures; B26 Tim Davis/Photo Researchers; B28 (li) Michael Giannechini/Photo Researchers; B28-B29 (bg) J.A. Kravlis/Masterfile; B28-B29 (ci) Ted Kerasote/Photo Researchers; B29 (ti) Superstock; B29 (bi) Mitsuaki Iwago/Minden Pictures; B30 (tl) David Muench Photography, Inc.; B30 (tr), (bl) Barry L. Runk/Grant Heilman Photography; B30 (br) David Muench Photography, Inc.; B32 Superstock; B34 (tli) V.P. Weinland/Photo Researchers; B34 (tri) Parviz M. Pour/Photo Researchers; B34-B35 (bi) Dembinsky Photo Associates; B34-B35 (bg) Larry Ditto/Bruce Coleman, Inc.; B35 (li) Tom McHugh/Photo Researchers; B35 (ri) Tom & Pat Leeson/Photo Researchers; B36-B37 Woods, Michael J./NGS Image Collection; B39 Bruce Coleman, Inc.; B40 (both) Joe McDonald/McDonald Wildlife Photography; B42 (bg) Stuart Westmorland/Stone; B42 (li) Roger Bickel/New England Stock Photo; B42 (ri) Bruce Coleman, Inc.; B43 (l) Kevin Schafer/Peter Arnold, Inc.; B43 (r) Mitsuaki Iwago/Minden Pictures; B44 (t) John Shaw/Bruce Coleman, Inc.; B44 (c) Hal H. Harrison/Photo Researchers; B44 (b) Wayne Lankinen/Bruce Coleman, Inc.; B45 (t) M. & C. Photography/Peter Arnold, Inc.; B45 (b) William Townsend/Photo Researchers; B46 (t) Vince Streano/The Stock Market; B46-B47 Ralph Ginzburg/Peter Arnold, Inc.; B48 Bryan & Cherry Alexander/Masterfile; B50 (t) Tim Davis/Photo Researchers; B50 (bl) Johnny Johnson/Tony Stone Images; B50 (br) Malcolm Boulton/Photo Researchers; B51 Tom McHugh/Photo Researchers; B52-B53 Ted Schiffman/Peter Arnold, Inc; B52 Roy Toft/Tom Stack & Associates; B54 Gunter Ziesler/Peter Arnold, Inc.; B55 (t) Doug Cheeseman/Peter Arnold, Inc.; B55 (b) Bonnie Kamin/PhotoEdit; B56 (i) Louisiana State University Chemistry Library Website; B56 Meckes/Ottawa/Photo Researchers; B60-B61 Craig Tuttle/The Stock Market; B61 (t) Jake Rajs/Stone; B61 (b) Earth Satellite Corporation/Science Photo Library/Photo Researchers; B62 Chromosohm/Sohm/Stone; B64 (t) David Muench Photography, Inc.; B64 (b) Gary Braasch/Stone; B65 (tl) Superstock; B65 (tr) Steve Kaufman/Peter Arnold, Inc.; B65 (bl) Joseph Van Os/The Image Bank; B65 (br) Colin Prior/Stone; B66 Wolfgang Kaehler Photography; B66 (i) Mark Moffett/Minden Pictures; B67 Superstock; B67 (i) Roger Bickel/New England Stock Photo; B68 David Muench Photography, Inc.; B68 (i) William Manning/The Stock Market; B69 Darrell Gulin/Stone; B69 (i) T. Eggers/The Stock Market; B70 David Muench Photography, Inc.; B70 (i) Joseph Van Os/The Image Bank; B71 Carr Clifton/Minden Pictures; B71 (i) Kennan Ward Photography; B72 (l) Nicholas DeVore, III/Bruce Coleman, Inc.; B72 (r) Tui De Roy/Minden Pictures; B74 Stan Osolinski/The Stock Market; B80 (t) Jim Brandenburg/Minden Pictures; B80 (b) David Muench Photography, Inc.; B82 (t) © Corel; B82-B83 Manfred Kage/Peter Arnold; B83 (t) NASA GSFC/Science Photo Library/Photo Researchers; B83 (bi) Pete Saloutos/The Stock Market; B84 (t) Romberg Tiburon Center; B84 (b) Emory Kristof/NGS Image Collection; B86 Jim Bradenburg/Minden Pictures; B88-B89 Gary Brettnacher/Stone; B89 (t) Jonathan Wallen; B89 (b) Argus Fotoarchiv/Peter Arnold, Inc.; B90 Frans Lanting/Minden Pictures; B92 Runk/Schoenberger/Grant Heilman Photography; B93 (t) Kennan Ward Photography; B93 (b) Ed Reschke/Peter Arnold, Inc.; B94 (t) Larry Nielsen/Peter Arnold, Inc.; B94 (c) John Marshall/Stone; B94 (b) Jeff & Alexa Henry/Peter Arnold, Inc.; B96 Art Wolfe/Stone; B98 Mark E. Gibson; B98 (i) Dr. E.R. Degginger/Color-Pic; B99 J.H. Robinson/Photo Researchers; B100 Francois Gohier/Photo Researchers; B101 Tony Arruza/Bruce Coleman, Inc.; B104 (c) Jim Corwin/Stone; B106 Tim Davis/Photo Researchers; B110 Mark E. Gibson; B111 (l) Bernard Boutrit/Woodfin Camp & Associates; B111 (r) Bill Tiernan/The Virginian-Pilot; B112 Courtesy of Atlanta Botanical Gardens; B112 (i) Kenneth Murray/Photo Researchers; B114 John Hyde/Bruce Coleman, Inc.; B114 (tli) Superstock; B114 (tri) Tom Bean/The Stock Market; B116 Centre For Ecological Studies; B116 (i) E. Hanumantha/Photo Researchers; B120 (t) Bill M. Campbell, MD; B120 (b) Graeme Teague Photography;

Unit C
Unit C Opener (fg) Eric & David Hosking/Photo Researchers; (bg) Bios (Klein-Hubert)/Peter Arnold, Inc.; C2-C3 Roger Werth/Woodfin Camp & Associates; C3 (t) John Livzey/Stone; C3 (b) Royal Osservatory, Edinburgh/Science Photo Library/Photo Researchers; C4 Tom Bean/Tom & Susan Bean, Inc.; C6 (tl) Helen Paraskevas; C6 (tr) Tom Bean/Tom & Susan Bean, Inc.; C6 (bi) Mark E. Gibson; C6-C7 Eric Neurath/Stock, Boston; C7 (t) NASA Photo/Grant Heilman Photography; C7 (b) Digital Visual Library/US Army Corps of Engineers; C8 (both) Mark E. Gibson; C9 M.T. O'Keefe/Bruce Coleman, Inc.; C10-C11 Michael Collier/Stock, Boston; C12 Soames Summerhays/Photo Researchers; C16 G. Gualco/Bruce Coleman, Inc.; C17 (t) Gregory G. Dimijian/Photo Researchers; C17 (c) Krafft/Explorer/Science Source/Photo Researchers; C17 (b) Tom & Pat Leeson/Photo Researchers; C18 UPI/Corbis-Bettmann; C20 M.P.L. Fogden/Bruce Coleman, Inc.; C23 Tom Bean/Tom & Susan Bean, Inc.; C24 A. J. Copley/Visuals Unlimited; C25 (t) R.T. Nowitz/Photo Researchers; C26 NASA; C27 (t) Walter H. F. Smith & David T. Sandwell/NOAA National Data Centers; C27 (b) David Young-Wolff/PhotoEdit; C28 (i) Santa Fabio/Black Star/Harcourt; C28 Tom Bean/Tom & Susan Bean, Inc.; C31 (l) Dr. E. R. Degginger/Color-Pic; C31 (r) Joyce Photographics/Photo Researchers; C32-C33 Dan Suzio/Photo Researchers; C33 (tl) Sam Ogden/Science Photo Library/Photo Researchers; C33 (br) Breck P. Kent/Earth Scenes; C34 The Natural History Museum, London; C36 (tl) Dr. E.R. Degginger/Color-Pic; C36 (bl) E.R. Degginger/Bruce Coleman, Inc.; C36 (bc) Joy Spurr/Bruce Coleman, Inc.; C36 (br), (c1) Dr. E.R. Degginger/Color-Pic; C37 (c2), (c3) E.R. Degginger/Bruce Coleman, Inc.; C37 (c5), (c6), (c8) Dr. E.R. Degginger/Color-Pic; C37 (c9) Mark A. Schneider/Dembinsky Photo Associates; C37 (c10) Dr. E.R. Degginger/Bruce Coleman, Inc.; C38 (tl) Dr. E.R. Degginger/Color-Pic; C38 (cl) Biophoto Associates/Photo Researchers; C38 (cr) Andy Sacks/Stone; C38 (bl) Dr. E.R. Degginger/Color-Pic; C38 (br) B. Daemmrich/The Image Works; C40 Joe McDonald/Bruce Coleman, Inc.; C42 (t) Dr. E.R. Degginger/Color-Pic; C42 (b) Phillip Hayson/Photo Researchers; C43 (tl), (tcl) Dr. E.R. Degginger/Color-P

(tcr) Breck P. Kent/Earth Scenes; C43 (tr) Robert Pettit/Dembinsky Photo Associates; C43 (b) Martha McBride/Unicorn Stock Photos; C44, C45 (tl), (tcl), (tcr), (tr) Dr. E.R. Degginger/Color-Pic; C45 (b) David Bassett/Stone; C46 (t) G. R. Roberts Photo Library; C46 (b), C46-C47, C47 Dr. E.R. Degginger/Color-Pic; C48 Tom Till/Auscape; C50, C51 (t), (b), C52 (l), (r), C52-C53 Dr. E.R. Degginger/Color-Pic; C54 James P. Blair & Victor Boswell/NGS Image Collection; C55 Mark Richards/Photo Edit; C56 Stuart McCall/Tony Stone Images; C56 (i) Photo Courtesy of Mrs. Alma G. Gipson; C60-C61 Bob Abraham/The Stock Market; C61 (t) NASA/The Stock Market; C61 (b) Stan Osolinski/The Stock Market; C62 (b) Warren Faidley/International Stock Photography; C64 (l) Everett Johnson/Stone; C64 (r) Warren Faidley/International Stock Photography; C65 (bg) Orion/International Stock Photography; C65 (bli) M. Antman/The Image Works; C65 (bri) Dr. E. R. Degginger/Color-Pic; C66 (t) David M. Grossman/Photo Researchers; C66 (b) Mark Stephenson/Westlight; C68 (l) Dan Sudia/Photo Researchers; C68 (tr) Kent Wood/Photo Researchers; C68 (cr) Kevin Schafer/Peter Arnold, Inc.; C68 (br) Gary Meszaros/Dembinsky Photo Associates; C74 Larry Mishkar/Dembinsky Photo Associates; C78 (b) Richard Brown/Stone; C80 (l) Tom Till; C80 (c) Blaine Harrington III/The Stock Market; C80 (r) Coco McCoy/Rainbow; C81 (t) Randy Ury/The Stock Market; C81 (c) Larry Cameron/Photo Researchers; C81 (b) Jeff Greenberg/Photo Researchers; C82 (cl) Ron Sefton/Bruce Coleman, Inc.; C82 (cl) Fritz Prenzel/Peter Arnold, Inc.; C82 (c) John Lawrence/Stone; C82 (cr) Marcello Bertinetti/Photo Researchers; C82 (r) Jose Fuste Raga/The Stock Market; C83 Paul Sequeira/Photo Researchers; C83 (l) Joe Sohm/Chromosomm/Photo Researchers; C84-C85 J. Richardson/Westlight; C86 (l) Brad Gaber/The Stock Market; C86-C87 A. Ramey/Woodfin Camp & Associates; C87 (l) NASA; C87 (r) Phil Degginger/Bruce Coleman, Inc.; C88 NASA/Goddard Space Flight Center/Science Photo Library/Photo Researchers; C88 (i) Eli Reichman/HRW; C92-C93 RKO Radio Pictures/Archive Photos; C93 (t) Thomas Abercrombiengs/NGS Image Collection; C93 (b) Clyde H. Smith/Peter Arnold, Inc.; C94 Sylvia Stevens; C97 (t) Timothy O' Keefe/Bruce Coleman, Inc.; C97 (b) Joseph J. Scherschel/NGS Image Collection; C97 (i) Carlos Lacamara/NGS Image Collection; C100 George D. Lepp/Photo Researchers; C102 Vince Cavataio/Pacific Stock; C103 (t) The Stock Market; C103 (ti) Michael P. Gadomski/Bruce Coleman, Inc.; C103 (bl) UPI/Corbis-Bettmann; C103 (br) Chip Porter/AllStock/PNI; C104 (t) Tony Arruza/Bruce Coleman, Inc.; C104 (r) Dr. Richard Legeckis/Science Photo Library/Photo Researchers; C105 George Marler/Bruce Coleman, Inc.; C106 (l), (r) John Elk/Stock, Boston; C110 (l) Brian Parker/Tom Stack & Associates; C110 (r) S.L. Craig, Jr./Bruce Coleman, Inc.; C111 (l) William E. Ferguson; C111 (r) Toms & Susan Bean, Inc.; C112 (t) Bruce Roberts/Photo Researchers; C112 (l) William Johnson/Stock, Boston; C112 (b) Wendel Metzer/Bruce Coleman, Inc.; C112-C113 Bob Daemmrich/Stock, Boston; C114 Michael Paris Photography; C116 (t) The Granger Collection, New York; C116 (bl) Corbis-Bettmann; C116 (br) The Granger Collection, New York; C116-C117 Eric Le Norcy-Bios/Peter Arnold, Inc.; C117 (tl) Naval Undersea Museum; C117 (tr) NASA/Science Photo Library/Photo Researchers; C117 (bl) Courtesy Smithsonian Diving Office/Photograph by Diane L. Nordeck; C117 (br) James P. Blair/NGS Image Collection; C118 (t) Woods Hole Oceanographic Institution; C118 (bl), (br), C119 (t) Emory Kristof/NGS Image Collection; C119 (c) Culver Pictures; C119 (b) Woods Hole Oceanographic Institution; C120 Allen Green/Photo Researchers; C121 Christian Vioujard/Gamma Liaison; C122 (t) PhotoDisc; C122 (b) Mike Price/Bruce Coleman, Inc.; C123 (t) Eric Freedman/Bruce Coleman, Inc.; C123 (b) David Woodfall/Stone; C124 Susan Lapides/Woodfin Camp & Associates; C124 (i) Emory Kristof/NGS Image Collection; C128 (t) Brownie Harris/The Stock Market; C128 (b) Adam Jones/Dembinsky Photo Associates;

Unit D
Unit D Opener (fg), (bg) Telegraph Colour Library/FPG International; D2-D3 Guodo Cozzi/Bruce Coleman, Inc.; D3 (l) Ray Pfortner/Peter Arnold, Inc.; D3 (r) Painting by Helmut Wimmer; D4 NASA; D6, D7 Frank Rossotto/StockTrek; D8 (t) Dennis Di Cico/Peter Arnold, Inc.; D8 (b) Frank Rossotto/StockTrek; D9, D10 (tr), (tl) NASA; D10 (ctl) Francois Gohier/Photo Researchers; D10 (ctr), (cbr) NASA; D10 (bl) Paul Stepan/Photo Researchers; D10 (br), D12 NASA; D14 (l) Peter Marbach; D14 (r) Jeff Greenberg/Unicorn Stock Photos; D15 Fred Habegger/Grant Heilman Photography; D18 NASA; D18 (Pluto) Dr. R. Albrecht, ESA/ESO Space Telescope European Coordinating Facility/NASA; D19 E. R. Degginger/Color-Pic; D20 NASA; D22 (t) The Granger Collection, New York; D22 (bl) Martha Cooper/Peter Arnold, Inc.; D22 (br) The Granger Collection, New York; D22, D23, D24 (bg) Science Photo Library/Photo Researchers; D23 (tl) Sovfoto/Eastfoto; D23 (tr) NASA; D23 (bl) Courtesy of AT&T Archives; D23 (br),D24, D25, D26-D27, D28, D29, D30, D33 NASA; D34-D35 Jerry Schad/Photo Researchers; D35 NASA; D36 StockTrek; D39 (t) Warren Faidley/International Stock Photography; D39 (c) Pekka Parviainen/Science Photo Library/Photo Researchers; D39 (b) Brian Atkinson/Stone; D40 (l) Rev. Ronald Royer/Science Photo Library/Photo Researchers; D40 (c) NASA; D40 (b) Hale Observatory/SS/Photo Researchers; D42 Wards Sci/Science Source/Photo Researchers; D46 John Chumack/Photo Researchers; D48 Andrea Dupree (Harvard-Smithsonian CfA), Ronald Gilliland (STScI), NASA and ESA; D49 Jeff Hester and Paul Scowen (Arizona State University), and NASA; D50 (l) NASA; D50 (r) Roger Ressmeyer/Corbis; D50 (b) Francois Gohier/Photo Researchers; D52 Bill Ross/Stone; D54 Fred Espenak/Science Photo Library/Photo Researchers; D55 Lynette Cook/Science Photo Library/Photo Researchers; D56 (t) Royal Observatory, Edinburgh/AATB/Science Photo Library/Photo Researchers; D56 (b) Robert Williams and the Hubble Deep Field Team (STScI) and NASA; D58 Dr. Robert Mallozzi of Universtiy of Alabama/Huntsville & NASA; D58-D59 Chris Cheadle/Stone; D59 Tony Freeman/PhotoEdit; D60 Courtesy of Julio Navarro/University of Victoria; D63 (l) Lynette Cook/Science Photo Library/Photo Researchers; D64 (t) Andre Jenny/Unicorn Stock Photos; D64 (b) Dennis Johnson/Folio.

Unit E
Unit E Opener (fg) Alvis Upitis/The Image Bank; (bg) Tim Crosby/Liaison International; E2 Jim Steinberg/Photo Researchers; E2-E3 Charles Krebs/The Stock Market; E3 H. Armstrong Roberts; E7 (b) NASA; E10 (bg) Ron Chapple/FPG International; E10 (bi) Dr. E. R. Degginger/Color-Pic; E12 Charles D. Winters/Photo Researchers; E14 Spencer Swanger/Tom Stack & Associates; E15 (c) Phil Degginger/Color-Pic; E16 Dr. E. R. Degginger/Color-Pic; E16-E17 Tom Pantages; E17 Tom Pantages; E18 Yoav Levy/Phototake; E19 Tom Pantages; E20 NASA; E23 (t) Tom Pantages; E23 (c) Yoav Levy/Phototake; E23 (b) Tom Pantages; E24 Horst Desterwinter/International Stock Photography; E25 (t) Dr. E.R. Degginger/Color-Pic; E25 (i) Norman O. Tomalin/Bruce Coleman, Inc.; E28 Joe Sohm/Photo Researchers; E29 (t) Doug Martin/Photo Researchers; E29 (b) Gary A. Conner/PhotoEdit; E30 (i) Glenn Photography; E30 Geoff Tompkinson/Science Photo Library/Photo Researchers; E34 Jan Taylor/Bruce Coleman, Inc.; E34-E35 Pete Saloutos/Stone; E35 Dr. E.R. Degginger/Color-Pic; E36 Superstock; E38 Lee Snider; E40 (t) J & L Weber/Peter Arnold, Inc.; E40 (b) Dr. E.R. Degginger/Color-Pic; E41 (Top to bottom) Joe Towers/The Stock Market; E41 (photo 2) Christopher S. Johnson/Stock, Boston; E41 (photo 4) Telegraph Colour Library/FPG International; E41 (photo 6) George Haling/Photo Researchers; E42 (li) Richard Laird/FPG International; E42 (b) Wesley Hitt/Stone; E44 Yoav Levy/Phototake; E48 Michael Monello/Julie A. Smith Photography; E50 (t) © PhotoDisc; E50 (bl) Mel Fisher Maritime Heritage Society, Inc.; E50-E51 (bg) Michigan Molecular Institute; E51 Michigan Molecular Institute; E52 (i) UPI/Corbis; E52 Mitch Kezar/Phototake; E56 (t) Sal Dimarco/Black Star/Harcourt; E56 (b) Courtesy of Jefferson Lab;

Unit F
Unit F Opener (fg) David Zaitz/Photonica; (bg) Stone; F2-F3 NASA/Photo Researchers; F3 (t) NASA/ Science Photo Library/Photo Researchers; F3 (b) Jean-Loup Charmet/Science Photo Library/Photo Researchers; F6 (l) Tony Duffy/Allsport Photography; F6 (c) Pascal Rondeau/Allsport Photography; F6 (r) Allsport Photography; F7 (tr) Spencer Grant/PhotoEdit; F7 (cr) Spencer Grant/PhotoEdit; F10 (b) David Young-Wolff/PhotoEdit; F12 (l) Myrleen Ferguson/PhotoEdit; F12 (r) Mark E. Gibson; F18 Superstock; F19 David Young-Wolff/PhotoEdit; F20 (b) Stephen Frisch/Stock, Boston; F21 (t) Novastock/PhotoEdit; F21 (ct) Tony Freeman/PhotoEdit; F21 (b) Tony Freeman/PhotoEdit; F24 NASA; F25 NASA; F26 (i) Dr. Ephraim Fischbach; F26 NASA; F30-F31 Chris Butler/Science Photo Library/Photo Researchers; F31 (t) Stephen Dalton/Photo Researchers; F31 (b) Mike Cooper/Allsport Photography; F32 (bi) E & P Bauer/Bruce Coleman, Inc.; F35 David Madison/Bruce Coleman, Inc.; F36 (i) Mark E. Gibson; F36-F37 Lee Foster/Bruce Coleman, Inc.; F38 (bl) Michael Newman/PhotoEdit; F40 A. C. Cooper LTD/Harcourt; F40-F41 Ed Degginger/Bruce Coleman, Inc.; F41 NASA; F42 Tony Freeman/PhotoEdit; F43 Mike Yamashita/The Stock Shop; F44 CP Picture Archive (Chuck Stoody); F46 Tom McHugh/Photo Researchers; F49 Erich Lessing/Art Resource, NY; F51 Scala/Art Resource, NY; F52-F53 Lawrence Livermore National Laboratory; F53 (t) Lawrence Livermore National Laboratory; F53 (b) Michael Rosenfeld/Stone; F54 NASA; F58-F59 NASA; F59 (b) M. Zhilin/M. Newman/Photo Researchers; F60 Jan Butchofsky-Houser/Dave G. Houser; F62 (l) Duomo Photography; F62 (c) William R. Sallaz/Duomo Photography; F62 (r) Steven E. Sutton/Duomo Photography; F63 (t) Gary Bigham/International Stock Photography; F63 (ti) Ken Gallard Photographics; F63 (bl) Stevn E. Sutton/Duomo Photography; F63 (br) Duomo Photography; F64 Greg L. Ryan & Sally Beyer/AllStock/PNI; F66 Peter Menzel; F68-F69 Phil Degginger/Bruce Coleman, Inc.; F69 (tr) Ontario Science Centre; F69 (br) Phil Degginger/Bruce Coleman, Inc.; F72 (tr) Michael J. Schimpf; F74 (bl) Stone; F76 (l) E.R. Degginger/Bruce Coleman, Inc.; F76 (c) E.R. Degginger/Bruce Coleman, Inc.; F76 (r) Tony Freeman/PhotoEdit; F78 (tl) Tim Beddow/Stone; F79 (tr) Danila G. Donadoni/Bruce Coleman, Inc.; F80 Norbert Wu/Stone; F82 Chuck O'Rear/H. Armstrong Roberts; F86-F87 Peter Cade/Stone; F88 (t) Michael Keller/The Stock Market; F88 (bl) The Granger Collection; F88 (bc) Kevin Collins/Visuals Unlimited; F88 (br) American Stock Photography; F89 (tl) Corbis-Bettmann; F89 (bl) Michael Nelson/FPG International; F89 (r) Lester Lefkowitz/The Stock Market; F90 Dr. Jean M. Bennett; F90 Diane Schiumo/Fundamental Photographs; F94-F95 Jeff Hunter/ The Image Bank; F95 (t) Science Photo Library/Photo Researchers; F95 (b) James King-Holmes/Science Photo Library/Photo Researchers; F95 (br) Ron Chapple/FPG International; F96 Cary Wolinsky/Stock, Boston/PNI; F98 Billy E. Barnes/PhotoEdit; F99 (tl) Gary Conner/PhotoEdit; F99 (tr) Bill Aron/PhotoEdit; F99 (br) Myrleen Ferguson/PhotoEdit; F100 (tl) Tony Freeman/PhotoEdit/PNI; F100 (b) David Young-Wolfe/PhotoEdit; F102 Jim McCrary/Stone; F104 (b) Wendell Metzen/Bruce Coleman, Inc.; F104 (i) Mark E. Gibson; F105 (tl) Keith Gunnar/Bruce Coleman, Inc.; F105 (tr) Mark Newman/Bruce Coleman, Inc.; F105 (c) Tony Freeman/PhotoEdit; F106 (l) Claus Militz/Okapia/Photo Researchers; F108 (bl) Myrleen Ferguson/PhotoEdit; F110 (t) Alan L. Detrick/Photo Researchers; F110 (b) Cameramann International; F111 (t) Nicholas de Vore III/Bruce Coleman, Inc.; F111 (c) Andrew Rakoczy/Bruce Coleman, Inc.; F111 (b) Glen Allison/Stone; F112 (tr) Lockheed Space and Missile Co., Inc.; F114 (t) © Corel; F114-F115 Mark E. Gibson; F115 (t) M.E. Rzucidlo/H. Armstrong Roberts; F115 (b) Andy Sacks/Stone; F116 (l) Drew Donovan Photography; F116 (r) Stan Ries/International Stock Photography; F120 (t) Courtesy of the National Science Center's Fort Discovery; F120 (b) Courtesy Associated Electric Cooperative, Inc.;

Health Handbook
R16 (c) Tony Freeman/PhotoEdit; R16 (br) David Young-Wolff/PhotoEdit; R17 (t) Myrleen Ferguson Cate/PhotoEdit; R17 (b) David Young-Wolff/PhotoEdit; R19 Tony Freeman/PhotoEdit; R45 (br) © PhotoDisc; R45 (bl) PhotoDisc;

All other photos by Harcourt photographer listed below, © Harcourt: Weronica Ankarorn, Victoria Bowen, Eric Camden, Digital Imaging Gorup, Charles Hodges, Ken Kinzie, Ed McDonald, Sheri O'Neal, Terry Sinclair & Quebecor Digital Imaging.

Illustration credits - Tim Alt B14-15, E15, E16-17, E39, E40, E41, F84; Scott Angle R44, R45; Art Staff C66; Paul Breeden A94; Lewis Calver A8, A9, A10, A11, A12, F78-79; Mike Dammer A57, A117, B85, C29, C57, C89, C125, D31, E53, F55, F117; John Dawson B38, B78-79; Eldon Doty A31, A57, A117, B57, C29, C125, D31, E53, F55, F117; Pat Foss A85, A117, B21, B57, B85, B117, E31, F27, F91; George Fryer C44, C46; Patrick Gnan F22, F65, F77, F78; Dale Gustafson B7, F64, F70-71, F72, F106; Nick Hall C102, C106; Tim Hayward A40; Jackie Heda A16, A17, A18, A19, A20, A25, A26; Inklink F20-21; Roger Kent A95, A100, A101, B44; Mike Lamble C16, C17, D6-7, D8, D9, D33, D55, F34-35, F48-49, F50; Ruth Lindsay B8-9, B92-93, B106-107; Lee MacLeod A27; Alan Male A79, A80, B43; MapQuest C75, C96; Janos Marffy C22, D15; Colin Newman B76-79, C110-111; Sebastian Quigley C64-65, C67, C72, C73, C75, C76, D14, D16-17, D38-39, D40-41, D46, D47, D48-49, F7; Rosie Sanders A102, A103; Mike Saunders A50, A51, C9, C10, C14-15, C18-19, C24 ,C31, C50-51, C98-99, C105, C118-119, F8; Andrew Shiff A31, B21, B117, C87, D61, E31, F27, F91; Steve Westin C42-43, C45; Beth Willert A24.